T0215815

Forschungs-/Entwicklungs-/Innovations-Management

Edited by
H. D. Bürgel (em.), Stuttgart, Germany
D. Grosse, Freiberg, Germany
C. Herstatt, Hamburg, Germany
H. Koller, Hamburg, Germany
C. Lüthje, Hamburg, Germany
M. G. Möhrle, Bremen, Germany

Die Reihe stellt aus integrierter Sicht von Betriebswirtschaft und Technik Arbeitsergebnisse auf den Gebieten Forschung, Entwicklung und Innovation vor. Die einzelnen Beiträge sollen dem wissenschaftlichen Fortschritt dienen und die Forderungender Praxis auf Umsetzbarkeit erfüllen.

Edited by

Professor Dr. Hans Dietmar Bürgel
(em.),
Universität Stuttgart

Professorin Dr. Diana Grosse vorm. de Pay,
Technische Universität Bergakademie Freiberg

Professor Dr. Cornelius Herstatt
Technische Universität
Hamburg-Harburg

Professor Dr. Hans Koller
Universität der Bundeswehr Hamburg

Professor Dr. Christian Lüthje
Technische Universität Hamburg-Harburg

Professor Dr. Martin G. Möhrle
Universität Bremen

Alexander Sänn

The Preference-Driven Lead User Method for New Product Development

A Comprehensive Way to Stimulate Innovations with Internal and External Sources

With a foreword by Univ.-Prof. Dr. rer. pol. habil. Daniel Baier

 Springer Gabler

Alexander Sänn
Bayreuth, Germany

Dissertation Brandenburg University of Technology Cottbus-Senftenberg, 2015

Forschungs-/Entwicklungs-/Innovations-Management
ISBN 978-3-658-17262-6   ISBN 978-3-658-17263-3   (eBook)
DOI 10.1007/978-3-658-17263-3

Library of Congress Control Number: 2017932120

Springer Gabler

Printed on acid-free paper

This Springer Gabler imprint is published by Springer Nature
The registered company is Springer Fachmedien Wiesbaden GmbH
The registered company address is: Abraham-Lincoln-Str. 46, 65189 Wiesbaden, Germany

Foreword

Lead user integration and conjoint analysis are widely known and commonly used methods in new product development that share the same goal – the improvement of a company's product offers – but have different advantages and disadvantages:

With conjoint analysis, on one side, a representative sample of customers is interviewed in order to measure the customers' preferences and willingness-to-pay for alternative improvements. By this, the companies' efforts are concentrated on improvements that a sufficiently large sample of customers will buy at a reasonable price after introduction. However, it is often said, that this approach also has its disadvantages. It tends to favor incremental improvements since the possible advantages of completely new improvements are beyond imagination and consequently esteem of the interviewed representative sample of customers. With lead user integration, on the other side, the companies focus on selected customers that are able to understand and anticipate the advantages of completely new improvements. The detection of such innovative customers is an important step of the approach. The preferences of these so-called lead users dominate the development of improvements and often lead to radical innovations. However, also this approach has its disadvantages. It is said to favor the development of so-niche products that the average customer will never buy at a reasonable price.

In order to eliminate the disadvantages of both approaches, Alexander Sänn proposes in his outstanding dissertation a combination of both methods, the so-called "preference-driven lead user method for new product development". The new approach mainly differs from the traditional lead user integration in so far that besides the lead users also a representative sample of customers is integrated in the new product development process in order to confront the lead users with the preferences of the average customers as early as possible. Both types of customers – average customers as well as lead users – are interviewed using conjoint analysis, the differences in their preferences are used for discussions.

Alexander Sänn has tested the new approach in various empirical studies. Two of these empirical applications are described in detail in his dissertation: One deals with an improvement of mountain bikes in the direction of electronic support, one deals with the improvement of security software for critical infrastructures. Both applications show that the new approach outperforms the traditional lead user method and conjoint analysis: The companies concentrate on more innovative improvements in comparison to following the recommendations of an ordinary conjoint analysis application and – at the same time – concentrate on improvements that tend to find a broader acceptance in the target segments.

The dissertation was accepted by the Brandenburg University of Technology Cottbus-Senftenberg in June 2015. Alexander Sänn received for his impressive work the grade "summa cum laude". I wish the work its well-deserved acceptance by the audience.

Bayreuth, February 2016

Univ.-Prof. Dr. rer. pol. habil. Daniel Baier

Preface

The present PhD thesis develops the Preference-Driven Lead User Method for new product development based on previous theoretical findings and practical experience. It was written during my academic time with Prof. Dr. Daniel Baier at the Brandenburg University of Technology Cottbus-Senftenberg and the University of Bayreuth. In detail, this thesis describes the challenging environment of new product development and highlights particular circumstances in Small- and Medium-sized Enterprises (SMEs). The new method is intended to solve selected challenges, e.g. idea evaluation and selection, by combining multiple activities within new product development to result in an integrated method.

The methodological development begins with a characterisation of the research environment and analyses the traditional lead user method and preference measurement of the new product development process. Particular challenges are discussed and aggregated based on examples from industrial practice. This thesis is intended to contribute to the findings by promoting the Preference-Driven Lead User Method with the background of the research question "Can the lead user method and preference measurement be combined to result in an integrated method for new product development?". The aggregation of idea generation, concept development, and concept evaluation is modelled within one comprehensive method against the actual sequential process.

Observations from the theoretical part point to multiple adjustments that can be made. The presented Preference-Driven Lead User Method makes use of the lead user method to stimulate ideation and links this to preference measurement while using a user-based recommendation algorithm to generate reliable acceptance data for every identified innovative contribution. This is developed as a combined approach and nested within the lead user method.

The new method is employed in the various application fields. The first example covers the field of mountain biking with 104 respondents and indicates heterogeneity in ratings of novelty and market potential. The second one covers industrial IT-security and aims to develop an intrusion prevention system for industrial networks with 246 respondents. The application showed promising results with an increased market potential and a decreased concept novelty. A further survey covered 311 respondents in the business fields of mechanical engineering industry, the automotive industry, and the field of market intelligence. The addressed business fields are presented separately with their specific market characteristics. The empirical investigation covers strengths and weaknesses of the lead user method per business field and evaluates the practical applicability of the new method. The results show that the Preference-Driven Lead User Method provides a benefit for future innovation projects.

Writing this thesis was only managable with the lasting background of my beloved family, my partner Kristin, and our daughter. I am so thankful for your support and for having you in my life!

This thesis would also not be possible without professional guidance. First, I thank Prof. Dr. Daniel Baier for his enduring support, supervision of the thesis, and for his guidance in academia. Second, I thank Prof. Dr. Magdalena Mißler-Behr for her review of this thesis and Prof. Dr. Uwe Meinberg for chairing the evaluation commission. I am much obliged to Dr. Alexandra Rese for her engagement in the commission and for our joint research in the field of Open Innovation. Further, I thank Prof. Dr. Peter Langendörfer and Prof. Dr. Rolf Kraemer for their support at IHP GmbH and our joint engagement in various research topics.

Dr. Sebastian Selka, Dr. Maria Marquardt, Dr. Eva Stüber, Dr. Ines Brusch, Jörgen Eimecke, Richard Bensch, Prof. Dr. Michael Brusch, Jana Krimmling, Stefan Lange and Dr. Said Esber deserve a great thank you. Thank you for all the fruitful talks and discussions, the motivational support, our friendship, and our time together. I also want to thank Project ESCI and its Lead Users. Especially, Dr. Joachim Müller deserves a great thank you for his support concerning the field of IT-Security in Energy Systems and for his trust in the ESCI system.

My team of undergraduate assistants, especially Felix Homfeldt, Philipp Schneider, Michael Keil, Martin Tietz, and Jens Fischer, deserves an outstanding appreciation as well as Beatrice Rich, Josefine Martha Pritschkoleit, Charlotte Irlen, Matthias Pantze, Jakob Levin, Zachery James Devlin, Kosta Shatrov, Franziska Kullak, and Vera Wessolek. Additional, thank you to all students whose scholar theses I were able to supervise. I appreciate your trust in letting me coach you!

Further, I wish all the best to the OUI Community and hope that this thesis will make a significant contribution and stimulates new research topics.

Table of Content

List of Tables

Table of Figures

List of Abbreviations

ACA	Adaptive Conjoint Analysis
AES	Advanced Encryption Standard
AHP	Analytic Hierarchical Process
ArbnErfG	Employee Inventions Act (Arbeitnehmererfindergesetz)
ASE	Adaptive Self-Explicated Measurement
BMBF	German Federal Ministry of Research and Education
BYOD	Bring Your Own Device
CA	Conjoint Analysis
CAD	Computer-Aided Design
CAP	Customer Active Paradigm
CAWI	Computer-Assisted Web Interview
CBC	Choice-Based Conjoint Analysis
CEO	Chief Executive Officer
CF	Collaborative Filtering
CIO	Chief Innovation Officer
CIP	Critical Infrastructure Protection
CoPS	Complex Products and Systems
CPS	Cyber Physical Systems
CRITIS	Critical Infrastructures (KRITIS)
CTO	Chief Technology Officer
DoS	Denial of Service (attack)
DTS	Driverless Transportation Systems
ECC	Elliptic Curve Cryptography
EEG	German Renewable Energy Sources Act
ESCI	Enhanced Security for Critical Infrastructures
F/OSS	Free/Open Source Software
FCHR	First Choice Hit Rate
FFE	Fuzzy Front End
FMCG	Fast Moving Consumer Good
GPL	General Public License
ICT	Information and Communication Technology
IEEE	Institute of Electrical and Electronics Engineers
IIT	Industrial Information Technology

IoT	Internet of Things
IP	Intellectual Property
IPS/IDS	Intrusion Prevention- / Intrusion Detection System
LAN	Local Area Network
LES	Leading Edge Status
LU	Lead User
MAP	Manufacturer Active Paradigm
MB	MegaByte
MITM	Man-in-the-Middle (Attack)
NDA	Non-Disclosure Agreement
NIH	Not-Invented-Here (Syndrome)
NPD	New Product Development
OEM	Original Equipment Manufacturers
OLS	Ordinary Least Squares Regression
OSI	Open Systems Interconnection Model
PASEA	Pre-Sorted Adaptive Self-Explicated Approach
PCPM	Paired Comparison–Based Preference Measurement
PLC	Programmable Logic Controller
QFD	Quality Function Deployment
R&D	Research and Development
RSS	Really Simple Syndication
SaaS	Software as a Service
SCADA	Supervisory Control and Data Acquisition
SEM	Self-Explicated Measurement
SIEM	Security-Information-and-Event-Management System
SME	Small- and Medium-sized Enterprise
SOTA	State of the art
TCA	Traditional Conjoint Analysis
UTM	Unified-Threat-Management System
VLSI	Very Large Scale Integration
VoC	Voice-of-the-Customer
WBUM	Web-Based Upscaling Method
WTP	Willingsness to pay
XML	Extensible Mark-up Language

1. Introduction

1.1. Overview

Innovation matters! Being innovative is a beneficial opportunity to step ahead of the state of the art and to discover new business models. Companies need to stimulate innovations to improve their internal processes, to establish new markets, and to differentiate their products in a mature market by addressing future customer needs. Nevertheless, being innovative is a challenging task and is generally assumed as demanding high investments to discover these new horizons. But, is it? Scientific literature in the field of innovation management relies on the formal process of new product development and offers a methodological variety to stimulate innovations – also with external input. Developing and commercialising innovations with users has already become relevant in science and practice. Still, being innovative is especially challenging in mature markets and demands major resources that are not available in every case and thus, this serious task may challenge a company's future existence. Therefore, literature and practice demand innovative methods to enable effective innovation management.

This thesis is intended to contribute to this demand by promoting the Preference-Driven Lead User Method with the background of the research question *"Can the lead user method and preference measurement be combined to result in an integrated method for new product development?"*. The aggregation of idea generation, concept development and concept evaluation is modelled within one comprehensive method against the actual sequential innovation management and promises key advantages for new product development with limited resources.

The author argues that it is possible to integrate market feedback in the early fuzzy front end of new product development before finalised concepts are developed. He reveals insights into the traditional lead user method in detail and to preference measurement in general. The author proposes the Preference-Driven Lead User Method and illustrates its application within two examples. He finalises his argumentation by drawing additional empirical research within three main German industries in order to challenge the traditional method.

Readers are introduced to the basic field of innovation management and grasp insights into challenges that lead to the research question.

They become familiar with the basic construct of "lead users" and learn about the ancestors of the traditional lead user method. They will know how to integrate lead users in new product development with respect to the importance of analogous markets and motivational aspects drawn by lead users and collaborating companies. The further methodological development of the lead user method is given to illustrate the state of the art and practical implications aside with its methodological challenges. This points to strengths and weaknesses of the lead user method. Readers will be further introduced to the field of preference measurement and will learn about selected methods to cover consumers' preferences. Various challenges of preference measurement for complex products and systems are further discussed. Readers will become familiar with the new Preference-Driven Lead User Method based on previous topics. Next, they will learn about expected benefits with practical examples from mountain biking and the application scenario of industrial IT security (IIT security) for CRITIS. Readers will also learn about practical circumstances to include lead users in the innovation process and get familiar with the evaluation of the proposed method in the German automotive industry, the field of mechanical engineering, and the field of market intelligence in Germany. This will guide readers to a detailed discussion and extra recommendations for further applications and future research.

Additionally, the VHB Jourqual 2.1 ranking of relevant literature in economic sciences was chosen to foster database analysis. The writing style is oriented on "Being Scheherazade: The Importance of Storytelling in Academic Writing" by Pollock/Bono (2013) and employs *italic type* to emphasise new terms and key aspects in the text. The structure within this thesis provides an overview and a concluding subchapter for each main chapter to address readers' convenience. The overview covers the basic content per chapter and shall manage readers' expectations. The conclusion expresses derived challenges in the theoretical part of the thesis next to implications in the empirical part of the thesis. This summarises relevant findings and contributions from the related chapter. The terms customer and consumer are used interchangeably, but depend on methodological chapter. Marketing research and market research are both summarised in the term of "market intelligence". The empirical application scenario was provided by IHP GmbH (Innovation for High Performance Microelectronics), Germany.

1.2. Innovation and Contribution

In general, innovations are new combinations of means and resources and trigger *creative destruction* (Schumpeter 1942). Schumpeter (1934) already provided the theory that the development of innovations is a necessary demand for companies to stimulate their growth and thus innovations are a major factor for economic growth in general. He also argues that the industrial and commercial life generates spontaneous and discontinuous changes but the wants (also known as needs and desires) of customers remain steady. Thus, spontaneous and discontinuous changes may be the drivers of innovation. Further, competition in innovation matters to stimulate innovation in quality, technology, supply management, and organisational management (Schumpeter 1942, p. 84). The prediction of successful innovations is a challenging task since historical data is used to support future management decisions and strategies. In contrast, capitalism is an evolutionary process and a form of permanent dynamic economic change. Thus, historical data may not be a reliable predictor (Schumpeter 1942) and an innovation will fail to meet economic expectations.

Modern approaches assume that an innovation will be successful if it is able to provide *benefits for the customer* (see Narver/Slater 1990), addresses their needs and is notably present at the market (as discussed by Sawhney et al. 2006). Consequently, literature argues to simply follow your customers' input (e.g. Jaworski/Kohli 1993 and Gruner/Homburg 2000). However, successful innovations require a proper understanding and translation for the customers' input to the company language (see e.g. Griffin/Hauser 1993 for an introduction to the voice-of-the-customer – VoC – propagation and Hauser et al. 2006). Traditionally, this provides only a snapshot on actual customers' demands and needs. This input will be mapped to predict future product requirements and may fail to generate successful innovations – e.g. in relation to an expected revenue. Such a failure of an innovation can be caused by further aspects, like communication and distribution, price, over-engineering, quality aspects, or a lack of internal support for the innovative product (see for example Hillenbrand 2007 for such a case in the automotive industry). Another major aspect is *trend forecasting* that is "… something of an art" (von Hippel 1986, p. 798).

A prominent example was experienced by Mercedes-Benz and its W140 S-Class that was built from 1990 in pre-series manufacturing to 1998. Its development started in 1981 and reflected customers' demands as well as social and economic trends. However, the S-Class faced a timing problem and was a symbol for social and economic trends that occurred in the 1980s, like economic improvement, opulence, and performance. These trends resulted in the biggest S-Class ever made. Unfortunately, trends changed when the W140 S-Class had its official market release in 1991. Trends then focused on a management of environmental resources and modesty. This occurred rapidly with the fall of the Berlin Wall, the Gulf War, and an unexpected economic recession. Thus, the best automobile in the world (Molitor 2007, p. 97) was not appropriate anymore and earned criticism for its philosophy, its size, and its fuel consumption. It was built by a world, for a world, that did not really exist anymore. On the one hand, this was a shock for Mercedes-Benz and the company suffered to regenerate from this development. On the other hand, this criticism was a turning point for Mercedes-Benz. As a result, the later A-Class was initiated. Molitor (2007) provides a retrospective view on the acceptance of the W140 and highlights the influence of changing trends to produce successful innovations.

Also, *R&D efficiency* is one of the important factors to stimulate successful products (see e.g. Cooper/Kleinschmidt 1987, George 2005 and Rese/Baier 2011 for a consistent importance of R&D quality and performance and Henard/Szymanski 2001 for necessary technological synergies). Surprisingly, only about 37% of surveyed companies said that they use formal R&D plans (see Jong/Marsili 2006). Figure 1 presents such a common staged R&D plan to stimulate successful innovations – the new product development process.

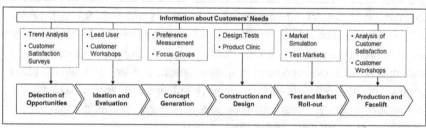

Figure 1 Stages in New Product Development
(Reference: Illustration in Reference to Baier 2000)

Small- and Medium-sized Enterprises (SMEs) face additional challenges like management difficulties, limited human resources, a lack of awareness for innovations, and changing development teams (see Freel 2005, Owens 2007, Cooper/Edgett 2008 and Parida et al. 2012).

The *historical background* of modern methods, schemes, and techniques in new product development is characterised by significant innovation flops. These flops showed that innovations bear high financial risks (see for example the case of the Ford Edsel in the 1950s that generated a loss of about $200 million – Hartley 1992) that bind most of the company's resources to new product development and commercialisation. Academic studies have shown that approximately 46% of all development resources go to unsuccessful projects, which do not reach commercialisation. Further 35% go to products that fail the commercialisation (see e.g. Cooper/Kleinschmidt 1987 and Cooper/Kleinschmidt 2000 for an update). Modern examples for failed innovations are e.g. Microsoft Windows Vista, Coca-Cola C2 or the Microsoft Kin smartphone product line (see Schneider/Hall 2011 for various examples and Sänn 2011).

Cooper/Dreher (2010) showed that especially product improvements – incremental innovations – became more relevant in their quantity. The authors noted that "…the nature of new-product development portfolios has shifted dramatically in the last 15 years away from bolder, larger and more innovative projects to smaller, lower-risk projects" (Cooper/Dreher 2010, p. 39). Projects for *radical innovations* – developing a new product and addressing a new market (see e.g. Lettl et al. 2006a and Lettl et al. 2006b) – are decreasing.

Risk becomes even more evident when dealing with *complex products*. Literature has not yet agreed on one universal definition but agrees that complexity will occur in multiple ways and at multiple levels with several indicators. Rycroft/Kash (1994) and Kash/Rycroft (2000) discussed complexity in relation to its technological impact besides the pure definition of complex products by their amount of attributes and attribute levels. Further views on complexity discuss for example the technological and market interdependencies (see e.g. Tidd 1997). Tidd (1997) presented a three staged approach of complexity that covers (1) a systemic view (amount of components), (2) multiple interactions (dependencies across different components etc.), and (3) non-decomposable relations (a product cannot be separated into its components without performance implications). Schmidt (2009) summarises that complex products are high-tech, cost expensive products and thus face a long-term buying process. They are interdependent, imply services, and thus provide a high benefit for the customer. In general, this is accompanied by multiple dependent components and a slow diffusion process (see Schmidt 2009, p. 97).

Today, this is accompanied by a *selection problem*. Modern web-technologies and methods of innovation management generate a vast amount of innovation opportunities and it is therefore the most important task to identify commercially promising ones (see e.g. Jensen et al. 2014) – finding a needle in a haystack. Against all odds, innovations are able to foster a competitive position within a market (as was noted by Debruyne et al. 2002).

Current research streams rely on external help to stimulate ideation and to compose superior product concepts. *Breakthrough innovations* are described as discontinuous innovations leading to advanced technological capabilities or enhanced product capabilities by combining knowledge from different fields (see e.g. von Hippel et al. 1999). It is analogue to *disruptive innovations* (see e.g. Christensen/Bower 1996 and Christensen/Euchner 2011) that point to simplicity and price advantages with lower performance, if necessary. This adopts the thought that everybody encounters problems with products of daily use and even a few users innovate or reveal their problems to the manufacturer. The *lead user method* is the state of the art technique to generate breakthrough innovations in the ideation and evaluation level (see figure 1). Preference measurement is the state of the art technique to generate product concepts. Previously gathered ideas are shifted to the level of concept generation.

However, an application of both levels in a sequential manner is not successful in every case (see e.g. von Hippel/DeMonaco 2013). Today, recent studies point to new opportunities, e.g. in social networks that provide collective evaluation of promising ideas (Lettl et al. 2009). Thus, the overall research question asks *"Can the lead user method and preference measurement be combined to result in an integrated method for new product development?"*. This thesis concentrates on this subject in order to contribute to successful product innovations. The author's assumption is that the integrated method will lead to new products with a decreased novelty but an increased market potential.

1.3. Structure of the Thesis

This thesis is structured in seven chapters (see figure 2).

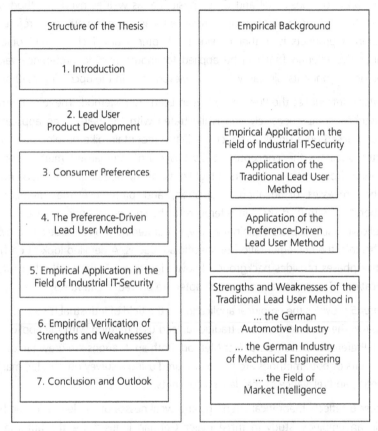

Figure 2 Structure of the Thesis

The theoretical part reflects the state of the art in lead user product development and customer preferences. *Chapter 2* explains the conceptual fundamentals and methodological ancestors of the lead user method. It introduces Eric von Hippel's thoughts based on cited references in von Hippel (1986), the definition of lead users, terms for lead user classification, methods of lead user identifications, and their motivational background. Furthermore, the lead user method and its adaptions are discussed. *Chapter 3* introduces the theoretical basement of consumer preferences and preference measurement for market intelligence. It explains the theoretical basement and highlights the concept of preference and utility modelling in detail. Methods for preference measurement are described briefly. Basic compositional and decompositional as well as hybrid methods are highlighted. A short evaluation of preference measurement methods in reference to complex products is further provided. Chapter 3 also includes the topic of collaborative filtering that can be applied to impute missing preference values based on respondents' similarity and is a relevant part in the upcoming method.

Chapter 4 introduces the Preference-Driven Lead User Method. The joint method is described comprehensively and is illustrated with the help of an application scenario based on a simple product – a fictive mountain bike development. This chapter discusses the conceptual basis based on summarised methodological challenges from previous chapters. It highlights the methodological adaption made to the processes of ideation, evaluation, and preference measurement in collaboration with lead users in reference to the sequential approach (see figure 1). Following subchapters explain the new, integrated method at large and provide insights into the implementation of specific adaptions. A detailed look at each of the five phases provides background knowledge to apply the method in future applications. The implications of this chapter are proved in chapter 5.

Chapter 5 provides the empirical application in the field of Industrial IT security and compares the application of the traditional lead user method with the application of the Preference-Driven Lead User Method. Both applications are drawn in detail. The results of both methods are then compared using a survey on novelty, market potential, and relevance of the derived concepts.

Chapter 6 reflects theoretical strengths and weaknesses of the lead user method within an empirical study in three major German industries – the automotive industry, mechanical engineering and market intelligence. The benefit of the new method will be discussed in reference to the results of the study.

Chapter 7 summarises the results of this thesis and provides a detailed discussion on methodological limitations, applicable adjustments and future research issues.

## 2.	Lead User Product Development

### 2.1.	Overview

It is crucial for businesses to generate innovations in order to promote growth and establish themselves competitively in a market through new product lines. This requires the generation of truly successful innovations and a certain trust by the senior management to the applied methods, which may result in an unexpected outcome. A huge amount of a company's resources and methodological knowledge is further required to transform theoretical knowledge from innovation management into practical application. These requirements may not be available within the field of SMEs. Thus, avoiding failed innovation processes and creating lucrative products become a more important topic in this area. Research has already shown that users themselves often drive market innovations. Von Hippel introduced the term 'lead user' in 1986 and fostered the lead user method. The lead user method incorporates specifically selected customers into the innovation process. This is especially apt in predicting future needs of basic customers and thereby finding optimal solutions to problems that have not been encountered yet. Lead users possess a unique set of qualities that allow them to identify future market trends and new problems before the bulk of ordinary customers do. Thus, lead users have extreme needs that will be relevant for the entire market in the future. However, the challenges within its applications are questionable and need to be highlighted such as the not-invented-here (NIH) syndrome, the local search bias, an unwanted transparency that is aside with the required trust. Therefore, this chapter is designed to present the influence of the lead user method for the development of innovations, to uncover challenges in the identification and integration of lead users, and to find recommendations for plans of action based on scientific literature and contributions from the field. Chapter 2.2 introduces the basic theory of lead users. It also reflects the lead user construct and its existence in multiple markets as well as motivational drivers. Chapter 2.3 reproduces detailed theory of the lead user method, introduces classification and identification, and highlights methodological adaptions that led to the state of the art. Chapter 2.4 highlights the practical relevance and shows recommendations for future applications. Chapter 2.5 summarises challenges that build the basis for the later introduced Preference-Driven Lead User Method.

2.2.   The Lead User Concept

2.2.1. Conceptual Fundamentals

Today, the research field of innovation management offers a wide range of tools, methods, and techniques to stimulate successful new products and service developments (see for a variety of methods and their practical relevance Cooper/Dreher 2010, Graner/Mißler-Behr 2013, and Rese et al. 2015). However, these tools and methods are placed within several stages of the innovation process. The *fuzzy front end* (FFE) of the innovation process describes the main tasks of ideation – the process of idea generation (Cooper/Dreher 2010) – and concept development (see Khurana/Rosenthal 1997, Kim/Wilemon 2002, and Verworn et al. 2008 for an overall view on innovation stages). The lead user method is placed in this fuzzy front end for ideation (see figure 1).

The traditional method for managing user integration in innovation management was the *Manufacturer Active Paradigm* (MAP), which was for example illustrated by von Hippel (1978) and was recently updated by Baldwin/von Hippel (2011). Schumpeter (1934) already mentioned that "It is, however, the producer who as a rule initiates economic change, and consumers are educated by him if necessary [...]" (Schumpeter 1934, p. 65). MAP in its Schumpeterian sense interprets the customer as a passive stakeholder, responding to new product developments by rejection or acceptance only while the company exclusively develops the innovation itself. The previously introduced example of the Mercedes S-Class (see Molitor 2007 and chapter 1.2) used to act in accordance with this paradigm. The customer is seen as an external idea provider, but the manufacturer was responsible for the new product development exclusively (cf. Robertson 1974, p. 331). Thus, external ideas are simply treated as a starting point for internal development processes. Bright (1969) raised early concerns about the neglected importance of analogous markets and encountered the need for a paradigm shift. For a long time this closed innovation model was standard practice (see Chesbrough 2003 for a short historical review on the closed innovation model leading to a paradigm shift).

The *Customer Active Paradigm* (CAP) understands the customer as a supplier of improvements, developing own solutions that are based on common products or even generating new and radical innovations (see von Hippel 1978 for an early application example from the scientific field). Thomke and von Hippel noted that the " *'need' information*" (Thomke/von Hippel 2002, p. 6) is a complex construct that cannot be gathered by MAP. Von Hippel (1994) characterises this need information as sticky in reason of the hard process to obtain it. Needs are hard to express by the customer and are challenging to translate, e.g. to the engineering-driven language of a company (see e.g. Griffin/Hauser 1993). Thus, he comments that conventional market intelligence techniques may fail and new methods for customer integration are required (see e.g. von Hippel 2001 for toolkits as a possible method to gain 'translated' ideas).

CAP reached its breakthrough with the concept of *open innovation* from a company's perspective (as introduced by Chesbrough 2003). Since then open innovation began to replace the dominating closed innovation approach in larger companies (see Chesbrough 2003 for first experiences and Huizingh 2011 for a historical review). Chesbrough/Brunswicker (2014) reported that today open innovation is well adopted in practice by large firms with about 78% usage ratio (n=125 from Europe and the United States). Besides the well-known advantages of open innovation, the method still includes restrictions for SMEs in its complexity and its demand for accessible resources like financial support, commitment, and knowledge (see Van de Vrande et al. 2009, Gassmann et al. 2010, DeMartino 2012, and Lasagni 2012 for challenges and demands for applications in SMEs).

The extended *CAP2* model places the user in the centre of commercialisation (as presented by Foxall/Tierney 1984). This is the state of the art paradigm to argue user integration, user innovation, and user entrepreneurship (see Shah/Tripsas 2007 for an early review and a detailed description and Bogers et al. 2010 for a review and future directions).

Today, there are several methods known to integrate the customer in the innovation process (see Creusen et al. 2013, Graner/Mißler-Behr 2013, and Rese et al. 2015 for an overview and brief definition of several methods applied in consumer and industrial goods markets). The *lead user method* chooses previously selected market participants who are included in the focus of the innovation process (cf. von Hippel 2005). This method has already become a quasi-standard to incorporate user innovations. Figure 3 illustrates MAP and CAP(2) with their level of the customer integration within the innovation process, their relative share of user-generated output, and the likelihood to generate successful products (as proved by Wind/Mahajan 1997, von Hippel 2005, and Cooper/Dreher 2010).

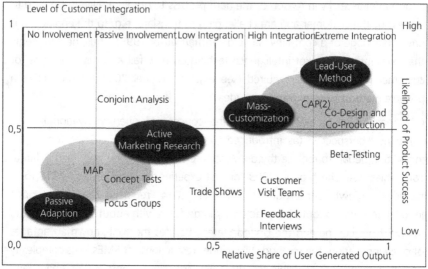

Figure 3 Positioning of the Lead User Method in Innovation Management
(Reference: Illustration in Reference to Wind/Mahjan 1997, Bilgram et al. 2008,
Cooper/Dreher 2010, and Horn/Brem 2013)

The market of sporting consumer goods was proven to be a valuable application field for this method. Several studies analysed this field to uncover *user-innovations* and lead user developments. Table 1 provides reference studies and shows that user innovations are quite frequent in this application field. For example, the empirical study by Franke et al. (2006) in the field of kite surfing revealed that about 30.9% of all 452 respondents modified their kite and may be lead users.

Table 1 Selected Studies of User Innovations in the Field of Sporting Goods

Application Example From Sporting Industry	Sample Size	Share of User Innovations	Reference Study
Canyoning	n=43	30.2%	Franke/Shah (2003)
Handicapped Cycling	n=19	26.3%	Franke/Shah (2003)
Kite Surfing	n=452	30.9%	Franke et al. (2006)
Mountain Biking	n=96	31.3%	Sänn/Baier (2012)
Outdoor Consumer Good	n=153	37.3%	Lüthje (2004)
Rodeo Kayaking	n=108	87.0%	Hienerth et al. (2014)
Sailplaning	n=87	41.4%	Franke/Shah (2003)
Snowboarding	n=48	18.2%	Franke/Shah (2003)

User innovations are also visible in business-to-business surroundings (see de Jong 2014 for an overview of related studies). De Jong/von Hippel (2009) observed up to 54% user innovations in a high-tech SME sample with 498 respondents from the Netherlands. This means that about 54% of the interviewed SMEs modified or invented products to solve their own needs. This is verified by findings from SMEs in Canada (see also de Jong 2014). Surprisingly, only 12.5% of corporate user innovators in the sample protected their modification by further intellectual properties (IP) rights. In general, literature agrees that approximately 30% of users within an application field are lead users (cf. Lüthje/Herstatt 2004). Table 2 summarises applications fields. The interpretation is analogue to table 1. For example, the empirical study of Lüthje (2003) in the business-to-business market (B2B) of surgical equipment revealed that about 22.0% of 261 respondents may be considered as being a lead user. About 37.8% of 197 respondents in the business-to-consumer market (B2C) of extreme sports equipments were lead users.

Table 2 Application Examples and Amount of User Innovations

(Reference: Illustration in Reference to von Hippel 2005, and Hienerth et al. 2014)

Examined Application Field for User Innovation	Market Type	Sample Size	Share of Lead Users	Reference Study
Extreme Sports Equipment	B2C	n=197	37.8%	Franke/Shah (2003)
Mountain Biking Equipment	B2C	n=291	19.2%	Lüthje et al. (2005)
Outdoor Consumer Good	B2C	n=153	9.8%	Lüthje (2004)
Apache Server OS	B2B	n=131	19.1%	Franke/von Hippel (2003)
CAD-Software	B2B	n=136	24.3%	Urban/von Hippel (1988)
OPAC System	B2B	n=102	26.0%	Morrison et al. (2000)
Pipe Hanger	B2B	n=74	36.0%	Herstatt/von Hippel (1992)
Surgical Equipment	B2B	n=261	22.0%	Lüthje (2003)

Recent studies suggest a stronger outsourcing of new product development activities to lead users (see for example Al-Zu'bi/Tsinopoulos 2013 for an application study on lead user benefits in a business-to-business surrounding). This is based on theory and is confirmed by a positive effect on cost reduction.

2.2.2. Methodological Ancestors

Eric von Hippel recognised the importance of user integration for a successful NPD process. Rothwell et al. (1974) have shown in the project SAPPHO (Scientific Activity Predictor from Patterns with Heuristic Origins) that the correct understanding of customer needs is essential for NPD. This laid the basement for other research streams (see e.g. Kutschke 2014). In reference, von Hippel noted that market intelligence analyses were typically not able to elaborate reliable input for the development of very novel products or "in product categories characterized by rapid change, such as 'high-technology' products" (von Hippel 1986, p. 791) and that user innovations were frequent in this example.

Von Hippel (1986) made use of *functional fixedness* as the main reason to argue why market intelligence analyses fail. Luchins (1942) proved that familiarity with complex solutions affects the solution competence for simple tasks negatively. The related experiment covered several problems all solvable by a complex procedure and presented these problem-solution combinations in succession to participants. Luchins (1942) showed that the participants were then blinded to solve a similar problem task by using a more direct and simple method. Their mind was fixed to stick to the complex solution strategy.

Duncker (1945) provided an extension of this statement. He observed that the use of a familiar object reduces the ability to find new solutions. This was confirmed by the studies of Birch/Rabinowitz (1951) and Adamson (1952). Therefore, users as problem-solvers are "... *constrained by their own real-world experience*" (von Hippel 1986, p. 791). Today, functional fixedness is still a basic problem in customer integration and can be seen as a staged problem task, e.g. while performing a problem-solving task and developing new products: (1) products and problems are characterised by their complexity and are dependent on embedded environmental constraints, (2) the user identifies existing usage patterns that fit the new product, (3) the new product's contribution to that existing usage pattern needs to be evaluated by the user, (4) new usage patterns that are enabled by the new product are abstracted, (5) the user evaluates the utility of the new product in the new usage pattern, and (6) the user does an estimation of how this is beneficial in contrast to alternative approaches that may be available to solve the problem. These six stages are very demanding for a customer. Therefore, functional fixedness leads to a fail of problem solving tasks (see von Hippel 1986).

This fosters any argumentation why *closed innovation fails* in generating truly novel innovations, especially in mature markets. Additional obstacles, which negatively impact market intelligence (cf. von Hippel 1986, p. 792), highlight that (1) there is no mechanism "... to identify all product attributes potentially relevant to a product category, especially attributes which are currently not present in any extent category member" (von Hippel 1986, p. 793), and (2) "[thus,] subjects are not well positioned to accurately evaluate novel product attributes or familiar product attributes which lie outside the range of their real-world experience" (von Hippel 1986, p. 793). In contrast, (3) the amount of attributes needs to be limited to be manageable and to fit in the frame of the development. Further, (4) the influence of the market analyst, who determines product functions, marketplace, and consumer leads to a misinterpretation of the gathered data.

Overall, these aspects express thoughts why market intelligence does not offer means of *going beyond the experience of the user*. Table 3 summarises von Hippel's argumentation to derive and prove user constraints.

Table 3 Referenced Literature on Functional Fixedness

(Reference: Illustration in Reference to von Hippel 1986)

Insights for Derivation of Lead User Existence	Reference Study
If somebody is familiar with a complicated problem-solving strategy, then it is unlikely that this person will use a simpler strategy when it is appropriate.	Luchins (1942)
Users are strongly blocked from using a familiar object (a product) in a novel way.	Duncker (1945), Birch/Rabinowitz (1951), Adamson (1952)
The aspect of functional fixedness is intensified by a recent usage of that familiar object.	Adamson/Taylor (1954)
The success of a research group in solving a new problem is dependent on whether solutions it has been used in the past will fit that new problem.	Allen/Marquis (1964)
An accurate understanding of user needs is essential for successful new product development.	Rothwell et al. (1974), Achilladelis et al. (1971)
Market intelligence for product perceptions and preferences is pertaining to functional fixedness.	Silk/Urban (1978), Shocker/Srinivasan (1979), Roberts/Urban (1985)

Von Hippel (1986) exemplary describes the usage of *focus groups* and preference measurement. He acknowledges that focus groups may be able to overcome these restrictions and may reveal unsolved problems with products. Nevertheless, this will be influenced by the analyst who is in charge to further pursue the unsolved problems or not. Also, the lack of a systematic approach is given. Thus, and with the basic background of the diffusion process (see table 4), von Hippel argues "... users whose present needs foreshadow general [market] demand exist" (von

Hippel 1986, p. 796). This means that these users "… are as constrained to the familiar as those of other users" (von Hippel 1986, p. 796), but they are ahead of the market and face a need that lies in the future for the bulk of the market. This means that lead users are also constrained by their functional fixedness, but live in the future and thus are not constrained from today's perspective.

Table 4 Referenced Literature on User Innovation and Diffusion
(Reference: Illustration in Reference to von Hippel 1986)

Insights for Derivation of Lead User Existence (Addresses Problem and Research Stream)	Reference Study
The link between innovation activity and expectation of an economic benefit exists. (User Innovation)	Schmookler (1966)
Innovations in industrial goods face a long-term diffusion process (75% of the empirical sample after 20 years). (Diffusion)	Mansfield (1968)
Acceptance of innovation typically develops successively through a society rather than affect all members simultaneously. (Diffusion)	Rogers/Shoemaker (1971)

Further, the greater a user's benefit from a solution is, the greater their effort will be (von Hippel 1986, p. 797). Users who will be able to obtain the highest *net benefit* (von Hippel 1986, p. 799; see subchapter 2.3.1) from the solution will be the ones who have devoted the most resources to understand the basic problem in both its complexity and surrounding. This leads to the possibility of providing reliable data for needs that are related to future conditions. Thus, Eric von Hippel describes lead users by using both terms. They face needs that will be in general in a marketplace – face them in advance – and will benefit from a solution for those.

The *derived systematic approach* is given by the lead user method. In a broad description they were named "inventive user" (von Hippel 1976, p. 234) and were notably observed in scientific application fields. Inventive users are the ancestors of the theoretical foundation of lead users (see for example von Hippel 1976).

2.2.3. Lead User Characteristics

Eric von Hippel introduced the term lead user to the field of innovation management in 1986. He described this user group as "… users whose present strong needs will become general in a marketplace for months or years in the future" (von Hippel 1986, p. 791).

Von Hippel (1986) defines lead users using two attributes:

1. "Lead users face needs that will be general in a marketplace – but face them months or years before the bulk of that marketplace encounters them", and

2. "Lead users are positioned to benefit significantly by obtaining a solution to those needs." (von Hippel 1986, p. 796).

Therefore, they "... *may innovate*" (von Hippel 2005, p. 22).

The 1st attribute describes the *potential for developing own innovations*. On the one hand, this attribute covers pioneering characteristics leading to an early adoption of new trends and market streams (as Lüthje/Herstatt 2004 and Schreier et al. 2007 discussed). This is correlated with a diffusion of needs and affect leading user groups before the bulk of the market (see von Hippel 1986 and van Eck et al. 2011 in relation to Rogers 1962, 2003 – adoption process and diffusion). On the other hand, lead users are able to experience new usage contexts and face requirements in advance of ordinary customers. This is because of their extreme usage contexts with no market-available solutions. Thus, they have to be inventive.

Consequently, the 2nd attribute describes the *individual benefits from an early solution*. This fosters lead users' motivation to perform innovative tasks, too, and to seek for cooperation with market-present companies to generate a solution that fits their needs (as can be seen in Lüthje/Herstatt 2004 and von Hippel 2005). That is why lead users are said to be "... sufficiently well qualified and motivated to make significant contributions to the development of new products or services" (Lüthje/Herstatt 2004, p. 554). Fundamentally and as a result of the given characteristics, lead users are not necessarily limited by functional fixedness (see table 3) and extend their creativity by not simply relying on their own use experience exclusively (as given by von Hippel 1986). In contrast to a random set of customers thinking within their ordinary real-world experiences and trying to integrate the innovation into a new (for them) and non-existing usage context, lead users are able to overcome this specific obstacle and think in their "future" experiences (summarised by Lüthje/Herstatt 2004 and subchapter 2.2.2). Following von Hippel's argumentation, they seem to think out of the box and to be beneficial sources of real novel product innovations (see von Hippel 1986 for first thoughts and von Hippel 2005 for later confirmation). These characteristics enable lead users to provide reliable information about their needs and problems and – in return –

allow the company to develop solutions for their customers based on these "future" needs (see von Hippel 2005). Thus, lead user contributions may foster long-term financial success (examined by Lilien et al. 2002).

Figure 4 positions lead users in the development phase of the product life cycle (cf. von Hippel 2005, p. 134) and includes the cycle of an "important market trend" (cf. von Hippel 2005, p. 22) as an interpretation sourced from Rogers (2003). This clarifies that lead users are in advance of early adopters with no solution for their need being currently market-available. Thus, they have to be inventors. Additional research extended both attributes by *implicit characteristics*, e.g. by opinion leadership, use experience with market-available products and product-related knowledge (see Urban/von Hippel 1988, Schreier/Prügl 2008, and Schuhmacher/Kuester 2012), and discuss the basic intrinsic motivation (see Lüthje 2000, Lüthje 2004, and Franke et al. 2006).

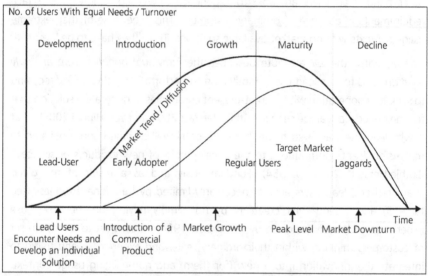

Figure 4 Positioning of the Lead User in the Product Life Cycle

(Reference: Illustration in Reference to von Hippel 1986, von Hippel et al. 1999, Rogers 2003, Morrison et al. 2004, and von Hippel 2005)

Table 5 summarises a compilation of lead user characteristics. The underlying work was done by Bilgram et al. (2008) and was further extended by recent studies. Exemplary, Kratzer/Lettl (2009) introduced the attribute of "*betweenness centrality*" (see Freeman 1979 for detailed research on this topic) that can be roughly described as an observation of a shortest path concept within a social network. This was proven to be beneficial for lead user specification and was observed among schoolchildren. They may be lead users when they act as a central (information) node within their network and possess a high betweenness centrality as a consequence (see Kratzer/Lettl 2009).

Table 5 Selected Additional Characteristics of Lead Users

(Reference: Illustration in Reference to Urban/von Hippel 1988, Lüthje 2000, Bilgram et al. 2008, pp. 426-428, Kratzer/Lettl 2009, Spann et al. 2009, Ozer 2009, Belz/Baumbach 2010, Füller et al. 2012, Faullant et al. 2012, and Poetz/Schreier 2012)

Being ahead of a market trend	Betweenness centrality	Creativity-relevant skills	Expertise	Extreme needs & circumstances of product use
Extrinsic motivation	Frequency and period of use	High expected benefit	Intrinsic motivation	Opinion leadership and word-of-mouth
Product related knowledge	Professional background or hobbyism	Speed of adoption	Strategic alignment	Use frequency of information sources
Development of modification	Dissatisfaction	Use(r) experience	User investment	Variance of use

Further, Spann et al. (2009) have tested constructs of "*opinion leadership*" and "*expertise*" to be sufficient lead user characteristics within a virtual stock market. They have shown that both characteristics led to notable differences in the output. The solely used attribute of a high-expected benefit (2nd point in the original definition) was shown to be unable to identify lead users accurately.

In contrast, Ozer (2009) showed that "*lead-usership*" and *expertise* would lead to different evaluation of new products, when both aspects were employed solely. Marchi et al. (2011) showed for online brand communities that the "strategic alignment with the brand identity" (Marchi et al. 2011, p. 358) needs to be specified as an additional lead user attribute to estimate the participant's "*level of innovativeness*". Füller et al. (2012) extended the then existing findings on beneficial lead user characteristics by "*creativity-relevant skills*" like tolerance for ambiguity, ability to delay gratification, independence of judgement, willingness to take risks and striving for excellence as well as "*task motivation*" (see Hennessy/Amabile 2010 and Füller et al. 2012 for detailed insights). Further, *"domain-specific skills"* like product knowledge, technical skills etc. are beneficial for a further development of user's innovation, e.g. to shift a basic idea to a market-available status. This supports the 1st aspect of the original definition of lead user characteristics (see Füller et al. 2012). Previously, Lüthje (2000) has illustrated that a *search for own developed ideas and concepts* can be employed exclusively to identify lead users and leads to sufficient results.

2.2.4. Analogous Markets

The *usage of analogies* is recommended to stimulate an increased novelty and radical innovations (see for example Dahl/Moreau 2002, von Hippel 2005, and Kalogerakis et al. 2010 for the positive impact of analogies). The benefits from including analogical thinking is well-known in innovation literature and was empirically proven to have significant influence to enhance the originality ($p<0.05$) and novelty ($p<0.05$) to the target market's solution (see Dahl/Moreau 2002 and Franke et al. 2014). Lüthje et al. (2005) have shown that in the case of mountain biking about 29% of the innovative concepts were conceived from an analogous field. Hienerth et al. (2007) confirmed these findings. Thus, the search for lead users should be extended to cover *"analogous markets"*.

Literature describes analogous markets as *different markets with similar needs* (see e.g. von Hippel 2005). Foreign markets may not have similar needs but are valuable to receive in-depth knowledge for particular aspects of development tasks, e.g. the knowledge of make-up artists for innovation focused on hygiene and dermal related needs (see for example the application by Herstatt et al. 2002).

Thus, ordinary customers from analogous markets can be lead users in the target market. This is beneficial to overcome the *local search bias* (see von Hippel 1994, Rosenkopf/Almeida 2003, Keinz/Prügl 2010, Poetz/Prügl 2010, Franke et al. 2014, and Herstatt et al. 2014, pp. 127-129 for positive examples) and restrain breakthrough innovations (see Schild et al. 2004 and Kalogerakis et al. 2010). The local search bias represents a form of functional fixedness and typically occurs in an organisational level (see Stuart/Podolny 1996).

The term "local" is described as a *contextual distance* (see Poetz/Prügl 2010). Ordinary customers in other markets might have similar characteristics, possess equal needs, but contribute to the innovation project by providing different perspectives on target market's problems from a distance. Thus, solutions from a foreign application field can be the starting point for a breakthrough innovation in the target market (as shown by Lüthje et al. 2005).

The lead user method operationalises this by *extending the search process* for lead users to cover multiple markets. For example, the search for knowledge about high performance brakes to generate a solution for the automotive industry was extended to cover the aviation industry (as shown by von Hippel 2005). Von Hippel et al. (1999) have shown the implementation of lead users from analogous markets in the application example of 3M. Table 6 illustrates selected studies.

Table 6 Selected Findings on Analogous Markets for Lead User Innovation

(Reference: Illustration in Reference to von Hippel 2005, Hienerth et al. 2007, Poetz/Prügl 2010, and Herstatt 2014)

Project (Application Field)	Target Market	Analogous Markets for Concept Generation	Reference Study
ABS Braking (B2B)	Automobile Manufacturers	Aircraft Industry	von Hippel (2005)
Automobile Parts (B2B/B2C)	Crush Protection for Passengers	Elevators, Doors and Gates	Hienerth et al. (2007)
Beverages (B2C)	Beer Cans	Biotech, PET, Toy Industry	Poetz/Prügl (2010)
Electronics (B2C)	Energy-Efficient Systems for Electronic Equipment	Cinema, Power Plants, Aerospace Industry	Poetz/Prügl (2010)
Escalators and Elevators (B2B)	Step Chain Systems	Conveyor Belts, Cable Car, Mining	Hienerth et al. (2007)
Food Industry (B2C)	Storage and Handling of Herbs and Spices	Pharmacy, Barkeeper	Poetz/Prügl (2010)
Infection Prevention (B2B)	Medical Staff	Microbiology, Hygiene Industry, Semi-Conductor	Herstatt et al. (2002)
Wireless Internet Products (B2B/B2C)	Informatics, IT Professionals	Military, Remote Diagnostics, Storm Chaser	Olson/Bakke (2004)
Medical Implants (B2B/B2C)	Medical Stents	Remedies	Herstatt (2014)
Medical Implants (B2B/B2C)	Hernioplasty	Design Studios, Automobile Seats, Clothing Industry	Herstatt (2014)
Medical Imaging (B2B)	Radiologists	Pattern Recognition, Semiconductors	von Hippel et al. (1999)
Surgical Drapes (B2B)	Doctors, Surgeons	Veterinary, Make-Up Artists	von Hippel et al. (1999)

Concerning analogous and foreign markets to foster a successful innovation is said to be the *major success factor for the lead user method* (see the argumentation by Lilien et al. 2002 and Hienerth et al. 2007). Modern literature builds on this aspect to introduce "Cross-Industry" innovation (e.g. Enkel/Gassmann 2010) with superior product attributes (derived from Kalogerakis et al. 2010 and Bruns 2013).

Lead users from analogous markets contribute significantly better to the concept workshop in terms of an increased novelty (p<0.05) and have no negative impact on the realisation of the concept (see Hienerth et al. 2007). This is known as the *"analogous market effect"* (cf. Franke et al. 2014, p. 1064). The effect was empirically proven to have a significant (p<0.05) positive impact on ideas' novelty, but also a significant (p<0.01) negative influence on ideas' usability in the target market (see Franke et al. 2014). This impact depends on the market distance in its contextual meaning. Contextually "near" markets perform better in usefulness (p<0.01) as where "far" markets perform better in novelty (p<0.05). Thus, this effect may overcome local search bias while increasing uncertainties.

Mistakenly, the construct of lead users is partially *confused with leading customers/ consumers* in terms of (1) adoption behaviour of new market-available products and (2) a high usage intensity of available products. Thus, it is essential that lead users and early adopters are separated before the motivational background is highlighted. Early adopters are characterised with the purchase in mind (see figure 4) and are generally considered to possess a higher income compared to a regular user. Lead users are characterised with the innovation in mind and are said to be pragmatic with low financial resources (see e.g. Slater/Narver 1998 and Lüthje et al. 2005). Recent studies have highlighted the relation between lead users and early adopters. A high degree of lead userness correlates with a fast adoptive behaviour, but not vice versa (cf. Schreier/Prügl 2008). Another fact to differentiate between both user groups is that "… lead users often serve during the commercialisation process as a reference group for early adopters, who [...] are reference groups for later adopters" (Droge et al. 2010, p. 71). Because of this background and their high-expected benefit from a solution for their need, they possess a personal interest in the joint development with established companies – a motivation that early adopters do not possess.

2.2.5. Motivational Background

Beneficial aspects, such as lower production costs, enhanced professional equipment available for a professional manufacturer, and the increased quality of the resulting product cause lead users to seek for collaboration with companies (cf.

von Hippel 2005). Intellectual property rights and joint commercialisation are from minor concern (cf. Harhoff et al. 2003). Lüthje (2004) shows that the expected financial benefit has no major influence on the decision to collaborate. Today, this attitude seems to change (see Franke et al. 2013 for this discussion).

This behaviour is known as *"free revealing"* (see von Hippel 1994 and von Hippel/von Krogh 2006 for its detailed origin). The lead user reveals their need information to the company and benefits from the solution fitting this need (as mentioned by von Hippel 1994). Free revealing was observed in several application examples akin to the companies 3M (see von Hippel et al. 1999) and Hilti (see Herstatt/von Hippel 1992). This is contrary to the basic economic drive (as noted by Henkel/von Hippel 2005) and leads to own investments and effort.

Today, this is in line with the CAP (see Baldwin/von Hippel 2011 and von Hippel 1978). Literature speaks in terms of user innovators and prosumers. The latter one illustrates the evolution from being a consumer to become a producer and consumer in the same market (an early reference to do-it-yourself prosumers can be found in Toffler 1980).

While the phenomenon of free revealing is a reasonable *answer to explain the motivation background* of lead users, this remains questionable for ordinary customers. Von Hippel/von Krogh (2003) discussed opportunities to stimulate free revealing behaviour for ordinary customers and presented the theoretical foundation by (1) the *private investment-model*, (2) the *collective action-model* and (3) the *private collective-model*. This leads to an *underlying motivation*, whereas: (see 1) private investments are made with the expectation of private rewards, e.g. by IP rights, (see 2) innovations are designed as a public good and are valued by compensational aspects, e.g. by reputation, and (see 3) innovations are "... created by private funding and then offered freely to all" (von Hippel/von Krogh 2003, p. 213). The authors examined this behaviour and its drivers in the case of *open source software projects* with users contributing their work – their private investment – to a collective innovation task.

Private investments are required to cover the costs of participation and contribution for each participant. Costs are described as any effort that is made by the contributor like their invested time. Participants in open source software projects are faced with the loss of proprietary rights and will not generate any direct earnings by the diffusion of the software. This excludes for example any additional income from direct product sales and is in line with the discussion of Kietzmann/Angell (2014) on the relationship between IP rights and creative customers. Lakhani/Wolf (2005) found that joy is the major driver in F/OSS projects. Literature argues with *intense, long-lasting, and sustainable cooperation* that is fostered by collective actions and will generate any additional benefit for a participant in an open source project. Furthermore, von Hippel and von Krogh explain that the loss of private investments will be balanced by indirect benefits from an *improved diffusion* and "… so increase an innovator's innovation-related profits through network effects" (cf. von Hippel/von Krogh 2003, p. 216).

This means that only a contributor can benefit from *subsequent network effects* in terms of profit. This leads to the assumption that contributors to the private-collective model will invest private effort if their *benefit from free revealing exceeds their benefit from private earnings*, e.g. by patents. This was described by von Hippel/von Krogh (2006) and is also applicable to lead users. The benefit from holding an innovation becomes a general calculation of effort and profit, e.g. by quantifying time investments to tinker a first solution vs. a low quality solution with drawbacks (see for example von Hippel et al. 2011). In sum, the benefit from free revealing an innovation needs to exceed the benefit from holding an innovation.

Benefit (free revealing) > Benefit (holding innovation)

For example, this expression may be valid if a patent is not valuable, but reputation is expected and additional rewards will be granted for free revealing and contributing the innovation to a joint development project. For lead users, the benefit from free revealing (e.g. reduced development effort, improved and available hardware and development tools) exceeds the sum of an own development (e.g. effort and the perceived low quality of their own solution). This is the basic driver for *private investments in collective development* and thus participation in open source projects can be beneficial. Contributions may serve as an opportunity for market-entrance to promote additional developments or to improve one's own reputation. Thus, private investments in collective development are expected to have beneficial influence on direct private investment models.

Additional literature on motivation found *learning and enjoyment* as private benefits for the contributor. They will further have a sense of ownership and thus will make valuable contributions to a joint community work. Similar observations were made in the case of the CipherChallange by Hall/Graham (2004). The social infrastructure to create a *sense of community*, the *desire to contribute to others' learning* and the benefit of *ones' individual learning* were identified as the main incentives to promote participation in collaborative work (see Hall/Graham 2004). This allows differentiating between hard and soft rewards. Hard rewards were found as *economic profit, access to non-public information* and one's own *career advancement*. Soft rewards were identified in terms of reputation and *satisfaction derived from the future solution for individual needs* (see Hall/Graham 2004). This is related to lead user characteristics, but is independent from the dichotomous lead user construct.

In pedagogy, this is known as *intrinsic motivation* (see e.g. Deci/Ryan 1993). Intrinsic motivation describes an interest-specific action, which has no external incentives for the condition and no external consequences. It explains the curiosity, exploration and interest in the immediate conditions of the environment. *Extrinsic motivation* occurs exclusively in connection with external motivational factors and is dependent from the external consequences as actions' result. Extrinsic motivated actions occur usually after orders. This happens in expectation of a positive consequence, like financial reward (see Deci/Ryan 1993).

The motivation of *sharing one's knowledge, dissatisfaction and a perceived major contribution to the community* were proven to have significant influence on one's willingness to participate ($p < 0.05$) in a company-hosted innovation contest (Wu/Sukoco 2010, Füller et al. 2010). Confirming observations were made in additional case studies that highlighted extrinsic rewards aside with *improving own skills* and being *addicted to the community* as drivers to positively influence one's willingness to participate (e.g. Brabham 2010 and Nambisan/Baron 2010). Surprisingly, Füller et al. (2010) observed that the willingness to help is negatively correlated to the willingness to participate for the case of medical equipment ($r=-0.36$, $p < 0.05$). However, *joy and hedonic interests* are intrinsic motivators to stimulate contributions with significant ($p < 0.001$) influence.

Table 7 provides an overview of intrinsic and extrinsic motivational factors and aggregates findings on users' motivation, free revealing and collaborative innovation. *Intrinsic and extrinsic motivational aspects* explain the benefit for a lead user who takes part in a lead user project. One may argue that they encounter a high-expected benefit – satisfaction – and thus participate in the innovation project solely based on intrinsic motivation. Current studies discuss intrinsic motivation as an exclusive fact for participation controversially (see e.g. Franke et al. 2013).

Table 7 Selected Aspects of Motivation and Willingness to Participate

(Reference: Illustration in Reference to Hall/Graham 2004, von Hippel 2005, Brabham 2010, Füller et al. 2010, Nambisan/Baron 2010, Wu/Sukoco 2010, Brabham 2012, Franke et al. 2013)

Motivational Category	Motivational Attributes to Stimulate Contributions	Reference Study
Extrinsic Motivation	Bounty and financial reward*	Franke et al. (2013)
	Access to new resources**	Nambisan/Baron (2010)
	Reputation, peer-recognition***	von Hippel/von Krogh (2003)
	Career advancement**	Hall/Graham (2004)
	Expand one's own ideation***	Füller et al. (2010)
Intrinsic Motivation, affective	Joy and hedonic interests***	Füller et al. (2010)
Intrinsic Motivation, norm-based	Knowledge sharing*	Wu/Sukoco (2010)
	Collective product improvement*	Wu/Sukoco (2010)
Intrinsic Motivation, rational	Individual learning***	Brabham (2010) Füller et al. (2010)
	(Dis-)Satisfaction*	Füller et al. (2010)
Significance: *$p<0.05$, **$p<0.01$, ***$p<0.001$		

From a *manufacturer's perspective*, the willingness to participate of lead users can be increased by providing appropriate surroundings (see Franke et al. 2010, Füller et al. 2010 and Wu/Sukoco 2010) like technical and supportive interfaces to allow contributions at low effort for the participant and to build trust and transparency. Fair IP policies (as discussed by Franke et al. 2013) help to improve the willingness to recommend the innovation task for a sufficient distribution in the community (see Dahlander/Magnusson 2008 and Wu/Sukoco 2010). Table 8 provides an overview of selected findings on *motivational factors from a manufacturer's perspective* to open corporate innovation projects. In general, collaboration with customers is beneficial. In detail, manufacturers can benefit from lead user implementation in terms of financial success and gain access to external R&D input (confirmed by Sanchez-Gonzalez et al. 2009). This is extended by positive observations of time-to-market, cost-to-market, fit-to-market and new-to-market aspects (see von Hippel 2005 and Schreier/Prügl 2008).

Further, Franke et al. (2010) examined that user collaboration leads to an increased *willingness to pay* (p<0.05) on the consumer's side and an increased preference fit (p<0.001) to address the consumers' benefit. Witell et al. (2011) showed that co-creation techniques like the lead user method outperform traditional market intelligence techniques in terms of originality of the developed new products.

Table 8 Performance Implications Provided by Selected Literature
(Reference: Illustration in Reference to Fuchs et al. 2011, Al-Zu'bi/Tsinopoulos 2012, Carbonell et al. 2012, Tsinopoulos/Al-Zu'bi 2012, Al-Zu'bi/Tsinopoulos 2013, and Butt 2014)

Application	Performance Implications for Future Application	Reference Study
Banking	Lead userness positively influences service advantages, service innovativeness*** and time-to-market, but has a negative effect on service novelty*.	Butt (2014)
British Manufacturers	Collaboration with lead users*** and suppliers** increase productivity. Lead user have more influence**.	Al-Zu'bi/ Tsinopoulos (2012)
European Manufacturers	Lead user*** and external product experts increase innovation rate. Lead user** enhances this further.	Tsinopoulos/ Al-Zu'bi (2012)
European Manufacturers	Lead user significantly decreases production costs**.	Al-Zu'bi/ Tsinopoulos (2013)
Product Manufacturers, R&D, CEOs	Direct and systematic interaction with experienced users and lead users*** increases innovation performances.	Fuchs et al. (2011)
Spanish Service Agencies With 75+ Employees	Lead userness positively affects service novelty***, service advantages*** and speed-to-market***, but a negative effect on market performance.	Carbonell et al. (2012)
Significance: *p<0.1, **p<0.05, ***p<0.01		

In contrast, Raasch et al. (2008) discovered circumstances that hinder user innovations. Technology maturity, technology complexity, satisfaction, innovation barriers – like unfair policies and IP regulations –, and the market concentration negatively influence the willingness to innovate. This leads to the assumption that immature and simple technologies support, an evolving dissatisfaction, and an absence of innovation barriers – like tight terms of use – favour user innovations. Raasch et al. (2008) further observed that a *hijacking adoption behaviour* by manufacturers may lead to rejection within the user community. The image of the manufacturer will be negatively affected although the new product may outperform the standard. The community may reject the new product and a sustainable commercialisation fails.

2.3.   The Lead User Method

2.3.1. The Traditional Lead User Method

The traditional lead user method adopts the basic lead user characterisation (in subchapter 2.2.3) and is motivated by observations made on a dominant role of user innovations in science (cf. von Hippel 1976). Von Hippel (1986) suggests an approach with *four phases to integrate lead users for NPD processes*. This is derived from his previous observations (see chapter 2.2.2) and is described by:

(1)   identifying an important market or technical trend,
(2)   identifying lead users who lead that trend,
(3)   analysing lead user data, and
(4)   projecting lead user data onto the general market of interest.

The (1) *identification* of *trends* is a core component of the company's activities to integrate lead users since it fosters the definition of lead user characteristics. Von Hippel (1986) differentiates between business-to-business and business-to-consumer markets, since industrial goods and consumer goods need different trend identification, evaluation, and are purchased under different conditions.

Von Hippel (1986) argues that the *clarity of trends depends on the market*. He mentions that the identification is simpler in markets for industrial goods than in markets for consumer goods. The value of a new product in industrial goods markets is measured mostly on *economic variables* bonding the underlying trend in close relation with the price of the future product and its anticipated benefits. Industrial goods are expected to be procured by pure economic decisions and thus trends are reflecting *efficiency and effectiveness*. For example, it has been made clear that the trend towards 'increased efficiency at lower prices' rises remarkably in the semiconductor business. In addition, it can be expected that this trend will continue over many years and fosters a long-term application field for NPD. Although, disruptive trends in this business area had been neglected for a long time (see e.g. Christensen/Euchner 2011 for the case of Intel Celeron processors).

In contrast, trends identified in consumer goods markets are seldom permanent and persistent and are also driven by non-economic decisions. Thus, the *prediction of a trend* seems to be almost impossible (see von Hippel 1986).

Table 9 illustrates von Hippel's (1986) differentiation between both markets. Von Hippel (1986) further suggests the Delphi method for the determination and evaluation of trends based on experts. Based on the results of the Delphi method it is the innovation manager's task to decide which trend should be pursued by the project team. Consequently, von Hippel (1986) recommends an improvement of methods for trend identification. In general, he noted that trend identification and trend evaluation are said to be in a "poor state of formal methods" in 1986 and remain *"something of an art"* (von Hippel 1986, p. 798).

Table 9 Patterns of Trend Identification in Industrial and Consumer Markets
(Reference: Illustration in reference to von Hippel 1986, p. 798)

Challenges of Trend Identification in Business-to-Business Markets	Challenges of Trend Identification in Business-to-Consumer Markets
Informal and accurate identification and assessment of trends is possible. Economic terms define the benefit. "Trends are inescapably clear to those in the industry." (von Hippel 1986, p. 798)	No reliable basement for assessment of trends is available. Lack of economic driven buying decisions. Unstable and inconsistent perceptions of trends over time by consumers.

The identification of trends is followed by the (2) *search for lead users* (as illustrated from the identification perspective). This aims to identify users "... at the leading edge of each identified trend in terms of related new products and process needs" (von Hippel 1986, p. 798) and "... who expect to obtain a relatively high *'net benefit'* from solutions to those needs" (von Hippel 1986, p. 798; see chapter 2.2). The net benefit can be operationalised in a business-to-business surrounding by the revenue (V) multiplied with the increased rate of profit (R) from applying a new solution minus expenses for the new solution (C) minus the margin using the old solution (D). This equals $B=VR-C-D$ (von Hippel 1986, p. 799).

The search for users at the *"leading edge"* in von Hippel's business-to-business setting can be done straight forward since a company's position on a specific trend is often well-known to industry experts. Further, manufacturers of industrial goods possess a familiar customer base that can be scanned easily for lead users, e.g. by the sales department. Moreover, the search for "high benefits" can be done by identifying a subset of leading edge subjects with massive R&D expenses among other criteria. Von Hippel introduces an example of the identified trend of "increasing density on chip surfaces" in 1986.

This usage pattern will be *exemplary* applied on manufacturers of memory chips. The leading edge position is derived from the integration of multiple transistors into a single chip at a very large scale integration level (VLSI; approximately hundreds of thousands of transistors onto a single chip; today this is negligible). Manufacturers at a VLSI level were at the cutting edge in these days and faced a strong need to increase the density of a single chip for reasons of efficiency and cost savings. Thus, they are in need of a higher profitability per single chip.

Additionally, von Hippel advises to identify users who are actively performing innovation tasks to solve problems of the leading edge. This was identified by investments in research and development practices in the semiconductor example. This addition implies that *high investments to research and development tasks* point to an unsolved problem and a high-expected benefit.

The next step is the (3) *analysis of the lead user data* to reveal user developed solutions and to gather substantive need statements. The accurate need statement and observed solutions are the foundation for the development of a new product and deliver a solution to a problem for the future market (von Hippel 1986, pp. 800-802). Thus, analysts need to find out in this *critical step* if lead user investments exist to solve a specific need. The analyst should interpret the derived insights into need statements and experience reports carefully.

The last step (4) *reflects the lead user needs to the overall market*. More important, the analyst must decide which needs are appropriate for the overall market, since *not all lead user needs will fit the future need of ordinary users and the mainstream market* (cf. von Hippel 1986, pp. 802-803). However, if buying decisions are not purely driven by economic characteristics, testing the solutions to determine a suitability for the future market seems to be very difficult (see von Hippel 1986). Thus, von Hippel (1986) suggests to check the acceptance of the solutions, for example, by prototyping. However, it must be guaranteed that future conditions will not change and that the user will have adequate time available to get to know the product well. Overall, new approaches are necessary for rapidly changing industries and for complex product development (von Hippel 1986, pp. 802-803).

Table 10 highlights recommendations for future methodological applications. The findings are based on the original article (see von Hippel 1986) and emphasise the relevance of analogous markets and the motivational background.

Table 10 Recommendations for Project Accomplishment by Eric von Hippel

(Reference: Illustration in Reference to von Hippel 1986, pp. 799-800)

Basic Advise	Recommendations for a Successful Lead User Method
Search for lead users in various markets	Lead users can also be found in the customer base of competitors in other markets (analogous markets, see 2.2.3).
Search for multiple lead users for problem solving	The company must not expect to find a lead user, who is responsible for the development of the entire product. Further, a lead user may only be able to solve a specific problem or to contribute to a particular component (using structured analyses).
Search for dissatisfied users alone is not sufficient	For lead users who have already solved a problem, a survey that is only directed to the unmet need leads to an unsuccessful identification.

2.3.2. Selected Methodological Adaptions

The integration of lead users to the NPD process has been highly discussed throughout academic research in general. An *analysis of multiple literature databases* (ScienceDirect, EBSCOhost, and Web of Science) revealed the actual relevance of the lead user method in form of annual publications. Figure 5 illustrates the amount of scientific publications. Overall, 812 filtered articles in relevant academic journals have been identified by September 30, 2014.

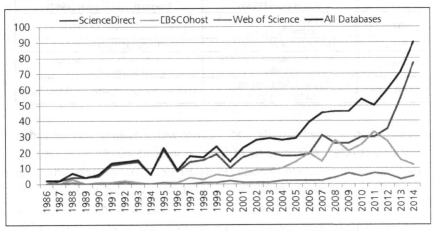

Figure 5 Annual Academic Publications from 1986 to 09/2014 (LU)

The keywords 'lead-user' and 'lead user' were used as search queries to scan the body of an article. Unfortunately, it could not be guaranteed for all databases that this excludes keywords in citations. Further, this includes related terms like 'somebody lead users'. Thus, the results per query were manually *filtered* with undergraduate assistance from off-topic posts, contributions that are not ranked in VHB JOURQUAL 2.1, and duplicates – articles in academic journals that are included in multiple databases or were presented twice in reason of the keywords.

However, *methodological adaptions* had been continuously made to the general process of the lead user method since 1986. An early *practical application* of the lead user method was done by Urban/von Hippel (1988) examining the effectiveness of the lead user method in the application field of computer-aided design systems (CAD). Herstatt/von Hippel (1992) did another practical application in the business-to-business surrounding in the company Hilti to develop a new generation of a pipe hanger. This case study addressed multiple decision makers – like architects, construction site managers, and facility managers – and often serves as reference for composing the process of lead user integration. Both application examples adapted the traditional lead user method to *fit practical needs* and extended the third step by a *lead user workshop*. Table 11 illustrates this.

Table 11 Traditional Lead User Integration and Application at Hilti
(Reference: Illustration in Reference to Herstatt/von Hippel 1992)

	von Hippel (1986)	Herstatt/von Hippel (1992)
Phase 1	Identification of market- and technology trends	Identification of trends
Phase 2	Identification of lead users	Identification of lead users
Phase 3	Analysis of lead user needs (and ideas)	Development of a concept via workshop setting
Phase 4	Applying lead user data to target market	Testing concepts appeal to ordinary users

This workshop is aimed to collaborate actively with intuitively selected and/or analytically selected lead users to perform an in-depth development task (cf. Urban/von Hippel 1988 and Herstatt/von Hippel 1992). For example, Urban/von Hippel (1988) chose five out of 38 identified lead users randomly and invited them to attend a workshop setting. Herstatt/von Hippel (1992) assessed two tests for the practitioners to check "… who seemed to be most appropriate to invite to join with Hilti engineers and other experts …" (Herstatt/von Hippel 1992, p. 218) instead of using a randomly selected lead user group. This selection process was conducted by the judgement of the interviewer and consisted of the two questions "Did the interviewer judge that the user could describe his experiences and ideas clearly?" and "Did the user seem to have a strong personal interest in the development of improved pipe-hanger systems?" (Herstatt/von Hippel 1992, p. 218). Hilti's pipe hanger systems framed the application field.

The *workshop setting* was designed using a 3-day schedule. The 1st day of the workshop discussed previous identified trends and problems of actual pipe hanger. Multiple thematic groups were formed in which a lead user on the 2nd day alternately worked. This separation in groups was done to avoid monotonous thinking. The results from the five groups were presented and analysed on the 3rd day. It were then the most promising concepts to be continued to work with until at the end of the 3rd day one final concept emerged (see Herstatt/von Hippel 1992 for a description of the applied lead user method and the workshop).

Von Hippel, Thomke, and Sonnack (1999) presented a further modified approach to apply the lead user method that is based on their experiences at 3M. Lilien, Morrison, Searls, Sonnack, and von Hippel (2002) referred to pyramiding (see subchapter 2.3.4 for details) for the identification process of lead users and refer to a similar methodological application at 3M. Table 12 illustrates this. The 1st phase was no longer characterised by the identification of a trend. Instead, it was reframed in order to *build the foundations* of the project and to define restrictions and requirements provided by the senior management (von Hippel et al. 1999). The determination of trends was shifted to the 2nd phase.

Table 12 Lead User Integration at 3M
(Reference: Illustration in Reference to von Hippel et al. 1999 and Lilien et al. 2002)

	von Hippel et al. (1999)	Lilien et al. (2002)
Phase 1	Project grounding	Goal generation and team foundation
Phase 2	Identification of trend	Trend research
Phase 3	Identification of lead users and their solutions	Lead user pyramid networking
Phase 4	Lead user workshop	Lead user workshop and idea improvement

The 3rd phase was characterised by identifying lead users and own developed concepts that will fit the company's "interests" (von Hippel et al. 1999, p. 52). The development of the concept can be found in the 4th phase that relies on the workshop and the determination of the fit to the target market. Lilien et al. (2002) pointed out that phase 4 of the lead user method employed at 3M was used "... to combine those [in phase 3 identified] ideas" (Lilien et al. 2002, p. 1055).

Olson/Bakke (2004) reframed this to a 5-phases method by *extracting concept testing* from the 4th phase to a 5th phase (see in reference to von Hippel 1986 and Herstatt/von Hippel 1992). Phase 4 was still reserved for the lead user workshop (see table 13). However, the process of von Hippel et al. (1999) was adopted by numerous authors (see e.g. Lüthje/Herstatt 2004 and Eisenberg 2011).

Table 13 Lead User Integration in Various Projects

(Reference: Illustration in Reference to Olson/Bakke 2004 and Eisenberg 2011)

	Olson/Bakke (2004)	Eisenberg (2011)
Phase 1	Planning the project	Preparation to launch the project
Phase 2	Determine key trend(s)	Identification of key trends and customer needs
Phase 3	Identify lead users	Exploration of lead user needs and solutions
Phase 4	Development of innovative ideas and product solutions	Improvement of solution concepts with lead users and experts
Phase 5	Concept testing	-

2.3.3. Selected Adaptions to Lead User Classification

The classification of lead users and non-lead users is a determination of a *dichotomous construct*, but empirical studies identified positive and negative effects on the quality of lead user contributions by the existence of single lead user characteristics (e.g. Poetz/Schreier 2012). Thus, the dichotomous construct of being a lead user or not being a lead user needs to be discussed.

This discussion was done by the *leading edge status* (LES) that was introduced as a continuous lead user scale (see Morisson et al. 2004). The LES is measured by constructs "benefits recognised early", "high level of benefits expected", "applications generation" and by a "perceived leading edge status" (cf. Morrison et al. 2004, p. 356). However, the classification and identification of leading edge users "...seems most severe in consumer goods fields where overall user populations appear to be 'unmanageably' large" (Schreier/Prügl 2008, p. 332).

Authors like Franke et al. (2006) and Schreier/Prügl (2008) coined the construct of *lead userness*. Franke et al. (2006) employed constructs of 'ahead of a trend', 'high benefit expected', 'technical expertise', and 'community-based resources' to determine lead userness. Its aim is to differentiate lead users from ordinary users by means of certain behavioural patterns (see Schreier/Prügl 2008). Thus, the latent construct of lead userness is defined by the relative trend position and the expected benefit from an innovation (see Franke et al. 2006 in relation with Schreier/Prügl 2008). Faullant et al. (2012) revealed a positive influence of 'use experience', product related knowledge and divergent thinking on lead userness. Overall, it remains arguable that these variables of the lead userness construct are consistent with functional fixedness (see Luchins 1942 and subchapter 2.2.2).

Table 14 gives an overview of statements used in literature to operationalise lead userness. *In fact, lead userness can be applied as a scale to evaluate the quality of a given user innovation, but not necessarily its market acceptance.* This is highly related to the construct of LES, but can be applied to emulate original lead users without an own idea development.

Table 14 Selected Operationalised Items to Measure Lead Userness
(Reference: Illustration in Reference to Schreier/Prügl 2008, Faullant et al. 2012, and Franke et al. 2014)

Dimensions	Exemplary Expression to Identify Lead Userness
Ahead of Market Trend	I am regarded as being on the 'cutting edge' in the field of [product]. I have a comprehensive knowledge of [product] available on the market. I have benefited significantly from an early adoption and use of [product]. I often get irritated by the lack of sophistication in [the product field]. I usually find out [information, products, and solutions] before others do.
High Expected Benefit	I am dissatisfied with the [reference product]. I am dissatisfied with the existing [equipment] offered on the market. I have improved and developed [a solution or product] myself. I have new needs which are not satisfied by [available solutions]. I have often noticed technical problems with [products]. In my opinion there are unresolved problems with [products].
Adoptive Behaviour	Annual spending on equipment Number of products owned Relative replacement rate of main equipment parts Relative time of adoption for a specific product

Franke et al. (2014) provide exemplary statements to *identify technical expertise*, e.g. "I can repair may own [equipment]", "I can help others to repair their [equipment]", "I am handy and enjoy tinkering" and "I can make technical changes to my [product]".

Related research to the construct of lead userness speaks of ancestors and consequences of lead userness. This illustrates the interaction between the lead userness and an actual development. Figure 6 shows an aggregated view.

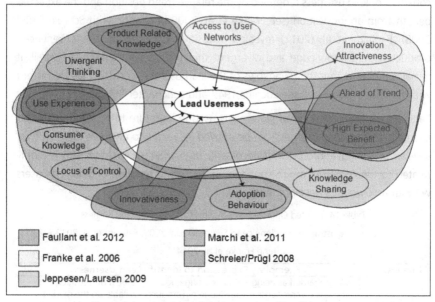

Figure 6 Antecedents and Consequences of Lead Userness
(Reference: Illustration in Reference to Franke et al. 2006, Schreier/Prügl 2008,
Jeppesen/Larsen 2009, and Faullant et al. 2012)

The distribution of the leading edge status and of lead userness within a given population could be observed as being bell-shaped (see Morrison et al. 1999) and was proven by Franke/Shah (2003) and Morrison et al. (2004) for business-to-consumer and business-to-business applications. Scientific research leads to the suggestion of *use the first third of the sample to identify lead users* (see for example Schreier/Prügl 2008 for a relation between a fast and heavy adoption behaviour, consumer expertise and use experience, locus of control and innovativeness, and one's lead userness). However, an own idea development serves as a sufficient requirement to identify user innovators and lead users in reference to standard literature (see e.g. Lüthje 2000 and von Hippel 2005).

As for now, it is proven that the presence of lead user(-ness) characteristics has a positive influence to an increased quantity and quality of user innovations and more attractive innovations (cf. Franke et al. 2006 and Schreier/Prügl 2008). In relation, Hienerth et al. (2007) emphasised that the source of a lead user's benefit – either from using or from selling the innovation – has a significant (p<0.01) influence on reproducibility of a contribution. Personal efforts to enhance their development are mainly driven by an expected benefit from a personal use (as given by Hienerth et al. 2014). Thus, the identification of lead users shall not be locked to the actual market, but be extended to further innovation activities in other fields with equal problem tasks (see chapter 2.2.4). Findings by Sänn/Baier (2012) and the discussion of Schuurmann et al. (2011) emphasise the challenge to determine lead users in the fuzzy front end of the NPD process.

In addition to the constructs of leading edge status and lead userness, Batinic et al. (2006) defined a *trend-setting scale* that is also used in practice to focus on early adopter characteristics and word of mouth to enhance the diffusion process and to stimulate market adoption of (lead) user innovations. The Trendsetting Questionnaire (TDS) does not necessarily imply an own idea development.

2.3.4. Selected Adaptions to Lead User Identification

The proportion of lead users within a sample is approximately 30%. Thus, the *identification of lead users* is a challenging and important task (see von Hippel 2005 and von Hippel et al. 2009) and thus various approaches have evolved.

Figure 7 visualises approaches to identify lead users that will be described below.

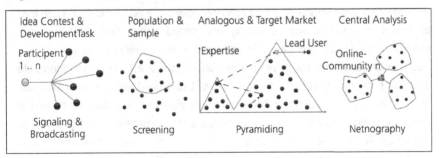

Figure 7 Illustrations and Comparisons of Identification Approaches
(Reference: Illustration in Reference to von Hippel 2005, von Hippel et al. 2009,
and Belz/Baumbach 2010)

The passive form of lead user identification is realised within the method of *signaling*. It represents an idea competition that will be announced within a company and if applicable to external recipients (as given by Tietz et al. 2006). The basic idea is that a lead user may be found behind a valuable contribution (see Jeppesen/Frederiksen 2006). Thus, the most important task to identify commercially promising ones (see e.g. Jensen et al. 2014).

Aside with signaling is the concept of *broadcasting* (e.g. Jeppesen/Lakhani 2010). Broadcasting promotes development challenges to external resources like a large crowd of potential problem solvers. This represents a modified perspective to problem solving, led to about 29% solve rate (see Lakhani 2006), and addresses the motivation of potential lead users to express own approaches and sharing knowledge within a community (as shown by Hienerth et al. 2007). The attention of the community members is obtained by e-mailing, internet advertising or short articles like contributions to journals or by releasing whitepapers. Broadcasting can be done, for example, in social networks, e.g. to take advantage of the small world phenomenon (see Watts et al. 2002 for a detailed description). The identification of lead user characteristics itself is done by a survey (as suggested by Tietz et al. 2006) and by self-selection among this crowd (see for example Martinez/Walton 2014 for a discussion on the wisdom of crowds). This fosters on pre-defined innovation tasks and thus ideas with a contextual distance may not be contributed.

The semi-active *screening* approach is a traditional method for lead user identification and is employed in the form of a questionnaire (e.g. described by Lüthje/Herstatt 2004). The main reasoning for identification based on screening is to generate a large sample size that probably covers the population. This can lead to hybrid approaches with signaling and broadcasting. Screening performs identification of lead users based on dedicated questions related to the employed lead user characteristics for determination (see chapter 2.2.1). The respondents are asked to self-evaluate reflecting constructs of LES or lead userness (see Morrison et al. 2000 and subchapter 2.3.3 for exemplary questions and statements) and to answer questions about their own innovations (see Urban/von Hippel 1988 and Sänn/Baier 2012 for an application example). In practice, respondents were given a set of exclamations to be ranked or to be rated. Further relevant statements may be "We consider it advisable to wait until others have pioneered new technology before adopting" and "We place great reliance on what other users tell us about the strengths/weaknesses of new technology" (see Morrison 1995, appendix I for the employed questionnaire) in a business-to-business surrounding. If the respondent would agree on both statements, then a LES would be negated. Respondents with the highest degree of fulfilment will be considered as being lead users (in line with von Hippel et al. 2009) and may be invited to the later lead user workshop. The screening approach is designed as a *parallel search process and is criticised for its inability to learn during the survey* (see von Hippel et al. 2009).

Another active assessment of lead user characteristics is possible when *pyramiding* is applied (see e.g. Lilien et al. 2002 as mentioned in table 13). Pyramiding sets the starting point to identify lead users among respondents in multiple markets and develops a network of respondents based on the theory of small worlds (see Milgram 1967). The interviewer asks if the respondent knows other people with specific expertise and stronger needs to fit the identification task (as given by von Hippel et al. 1999). This will be continued until a pre-defined level of lead user attributes is reached. Questions concern, for example, an individual development, a detailed report on a specific problem within a trend, and further items of lead userness (for discussions on this approach see von Hippel et al. 1999, Poetz/Prügl 2010, and Stockstrom et al. 2012). Thus, the pyramiding approach is designed as a *sequential search process* and will lead to respondents outside of the target market (as described by von Hippel et al. 2009 and Poetz/Prügl 2010). Fundamentally, the interview fosters on the questions *"What do you know?"* and *"Who you know?"* (Churchill et al. 2009, p. 82).

Netnography is a pro-active method for the analysis of online communities (based on definition by Kozinets 2002) and extends ethnographic research methods to internet communities. This approach uses information from social networks or online discussion boards to investigate the needs and the information seeking behaviour of consumers to reflect underlying trends and their market implications (see Belz/Baumbach 2010 for an application example). The product development team that relies on analysing community contributions carries out the evaluation. The netnography approach consists of four steps: (1) identify a suitable online community; (2) data collection and analysis; (3) interpretation of the data; and (4) inform the community members about the analysis to guarantee ethnically proper research. Belz/Baumbach (2010) employed a screening approach to verify lead user identification via netnography as an optional step (5). Additional research promotes an automatic identification process (see Pajo et al. 2013).

Market intelligence and innovation management can therefore rely on a toolbox for lead user identification that supplies the appropriate method depending on the requirements and conditions. The decision to employ a specific approach for lead user identification depends on the requirements and the environment of the application field. In practice, it is relevant to know whether the ideation task is done in an anonymous market or in a well-interconnected application field.

Pyramiding is recommended in well-connected communities (see Stockstrom et al. 2012) as they are supposed to exist in business-to-business markets (see von Hippel et al. 2009). Further, pyramiding allows deepening into analogue markets (as confirmed by e.g. Poetz/Prügl 2010). Von Hippel et al. (2009) showed that the pyramiding approach represents an efficient search method for the identification of lead users and outperforms the screening approach in specific cases with an average effort of only 28.4% in comparison to the screening approach. This result is based on simulated data. Stockstrom et al. (2012) support this by similar findings with an average effort of 31.0%. Effort is described by the number of contacts to identify the person with the highest level of a given attribute in a population (*efficiency*). This depends on the size of the population and leads to better results for traditional screening when it is performed in a small market (also in Stockstrom et al. 2012). In addition, findings also show a significant correlation between the target person's reputation (r=0.37, p<0.1) and pyramiding efficiency. Thus, if a community or market would have a role model, then pyramiding will be in favour of use. Moreover, pyramiding allows to respect important learning effects from each respondent by the project team and possesses benefits when used in poorly mapped search spaces (see Poetz/Prügl 2010). From the manufacturer's perspective pyramiding is highly interesting since the overall referrals per interview is about 0.96 according to the eight examined lead user studies by Poetz/Prügl (2010) and allows the company to build up a network of lead users. Better, lead users tend to provide significantly (p<0.01) more referrals into analogous markets than non-lead users. Contacts from analogous markets provide additional contacts to cross domain-specific borders (as proved by Poetz/Prügl 2010).

Screening is especially recommended for anonymous communities like in business-to-consumer markets, but with a known population (see Lüthje et al. 2005). In contrast to the advantages of pyramiding, considerations can be expressed that favour the screening approach because of the quantity of identified contributions. Lüthje/Herstatt (2004) remarked the quantity of contributions as an important criterion for the generation of ideas in the early phase of the innovation process (the fuzzy front end). Thus, the choice of an identification method depends whether the project focuses on idea gathering or on cherry picking. A resource-based perspective on this choice would also consider *effectiveness*. Screening requires a suitable survey distribution once, whereas pyramiding requires an in-depth interview each time. Literature has shown that "... the cost of the screener method per lead user identified was [...] US-$ 9,767" whereas "... the cost per lead user identified via the pyramiding procedure in this real world case was US-$ 1,500" (von Hippel et al. 2009, p. 1399). Summarised, von Hippel et al. (2009) acknowledged "... the relative efficiency of pyramiding will be contingent upon the presence of one or many specific conditions or factors" (von Hippel et al. 2009, p. 1404). In addition, today's web-technologies enable fast and cheap feedback (see Baier/Sänn 2015). Table 15 provides an overview to main objects and obstacles for lead user identification using an active method.

Table 15 Screening, Pyramiding and Netnography in a Nutshell
(Reference: Illustration in Reference to von Hippel et al. 2009, Stockstrom et al. 2012, and Belz/Baumbach 2010)

Main Objective of Identification	Advantages of the Employed Approach	Obstacles Revealed by Empirical Work	Estimation of Efficiency	References Study
Netnography - user-insights on expressed needs	Contributions without interference by project team, community ties	Minimum number of posts required, limited information about community	83% (5 out of 6 identified)	Belz/Baumbach (2010)
Pyramiding - qualitative ideation	'On the fly' learning, problem solvers from analogous markets	Unknown success criterion, measurement of thresholds, starting point	10% up to 65% (avg. 28% to 31%)	von Hippel et al. (2009), Poetz/Prügl (2010), Stockstrom et al. (2012)
Screening – quantitative ideation	Broad coverage	No learning effects, self-evaluation	100% (benchmark)	Lüthje (2000)

2.3.5. Selected Adaptions to the Methodological Scope

An adaption to the scope of lead user projects was provided by Henkel/Jung (2010) using a *Technology Push Lead User Concept*. This scope aims to identify future product concepts based on a given technology. In other terms, lead users develop problems that can be solved with an existing technology. Thus, they may act as a reference group for early adopters.

Sänn et al. (2013) derived an approach for the integration of lead users to foster parameterisation of research projects. The *Preference-Driven Lead User Intelligence* was especially applied for the usage in a research facility from the high technology sector that deals with increased complexity of the product and accelerated product life cycles. Thus, it addresses the challenge of dealing with restricted resources to perform innovative tasks, like a limited financial background, superior technical knowledge and multi-project team members that do not exclusively work on the lead user project. These circumstances occur, for example, in publicly funded and technological-driven research projects. Based on those observations, this approach aims at the generation of parameters for future technology developments in general, and not to generate fully developed product concepts with lead user integration. Following this, the innovation manager is centred as a provider of future parameters and product attributes in orientation the lead users. Thus, this scope may be applicable to foster future research tasks. Table 16 compares both approaches to the traditional lead user method.

Table 16 Original Lead User Method and Selected Methodological Adaptions
(Reference: Illustration in Reference to Henkel/Jung 2009 and Sänn et al. 2013)

	von Hippel (1986)	Henkel/Jung (2009)	Sänn et al. (2013)
Phase 1	Identification of market- and technology trends	Determines the relevant characteristics of the focal technology	Identification of experts, trends, and rudimental ideas
Phase 2	Identification of lead users	Search for trends that are furthered by these characteristics	Specification of lead user indicators and survey preparation
Phase 3	Analysis of lead user needs (and ideas)	Identify markets in which these trends matter	Lead user identification and market verification
Phase 4	Applying lead user data to target market	Identify lead users	Deriving parameters and concepts for development
Phase 5		Generation of application ideas	

2.4.   Lead Users in Practice

2.4.1. Applications and Recommendations

The lead user method was successfully applied for industrial goods and for consumer goods as well as in service development (e.g. Oliveira/von Hippel 2011 and de Jong 2014). Table 17 summarises selected application examples of lead user studies with derived *recommendations for further applications*. Common aspects like translating consumers' expressions to usable input are major challenges although previous literature like Griffin/Hauser (1993) has already pointed to this. An extended list with a focus on lead user identification and integration approaches is for example given in Gängl-Ehrenwerth et al. (2013, pp. 378-379).

Table 17 Selected Studies of Lead User Integration and Recommendations

Empirical Application	Recommendations for Future Application	Reference Study
CAD-software	Similar evaluative structures of lead users and non-lead users foreshadow a positive adoption.	Urban/von Hippel (1988)
Pipe hanger	The mixed workshop setting improves the overall teamwork process and creates a common language, e.g. for engineering and economics.	Herstatt/von Hippel (1992)
Home banking services	Lead users are not to confuse with early-adopters, thus identification should focus on own innovations that need to be understood by the team and not simply copied.	von Hippel/Riggs (1996)
Surgical equipment	Lead user integration demands a high level of commitment needed from team members and senior management, e.g. in time, company resources, and trust.	von Hippel et al. (1999)
Desktop PC/groupware application	Lack of time, resource pressures, unsupportive corporate culture, and disappointing prior experiences hinder the implementation of the method.	Olson/Bakke (2001)
Mechanical engineering	Key influencing factors are the time and intensity of involvement and the form of governance to make the method successful.	Enkel et al. (2005)
Sporting goods	Experts rely on solution information and consumers provide need information that need to be combined to fit users' needs.	Piller/ Walcher (2006)
Information & communication technologies	Team members need to be open-minded and should expect unexpected results or ways to be further followed.	Eisenberg (2011)

The application for CAD-Software development was shown by Urban/von Hippel (1988) and led to significant (p<0.01) differences reflecting the first choice hit rate (FCHR) to prefer the lead user concept with respect to alternative solutions available on the market. An additional performed constant sum measurement confirmed the preferred lead user concept ($p<0.1$) and showed that the non-lead user group would prefer the lead user concept, too. Overall, positive results were observed in the majority of published empirical applications. This highlights that the lead user method develops superior products, which are new to the market and outperform market available products (see Eisenberg 2011). Literature also shows *negative examples of lead user integration*.

Olson/Bakke (2001) employed the lead user method at a company named Cinet. In this case, the lead user integration went wrong since appropriate commitment by managers was absent and the workshop was completely dominated by the invited lead users and lacked in discussion. Thus, the results were biased. Cinet decided to stop the release of the lead user concept to the market. It cannot be said that the lead user method failed, but the *necessary support and management's commitment* were missing. The lead user method did not become a permanent tool in this enterprise.

Enkel et al. (2005) implemented the lead user method to a company called Swiss Engineering (in a business-to-business surrounding) to reduce innovation risks for a future development called Betty. This innovation process started with traditional market intelligence to develop a product that fits customer needs. Swiss Engineering changed its strategy to be fostered on the lead user method to reduce market uncertainties. This led to *resistance* within the company. The project team was not convinced of a beneficial integration of customers, but were forced to stick to the method. Further, Swiss Engineering employed the screening approach to identify lead users according to pre-defined lead user characteristics. In contrast to theory (see e.g. subchapter 2.2.4), they applied this identification only on selected reference customers and early-adopters from their own customer database (assumingly in reference to Herstatt/von Hippel 1992), but did not cover the population in their market. Leading customers were identified, but they lacked in technological expertise and were excluded from the innovation process. Overall, this led to later resistance from the customers' perspective. The before interviewed customers were asked a 2nd time about their preferences and revealed needs that required major changes. This unreliability led to *uncertainties* at the senior management and the project was stopped. Overall, the method was employed after the prototype has almost been finished. This led to insuperable challenges.

2.4.2. Performance Measurement

Studies on customer involvement in successful innovations have revealed the importance of lead user integration (see for example Lilien et al. 2002 and von Hippel 2005) and their contribution to successful new product development (as proved by e.g. Franke and Shah 2003 and Lüthje and Herstatt 2004). Early hints for the superiority of the lead user method from a practical perspective was given by Herstatt/von Hippel (1992) referring to a *shortened development cycle and decreased development costs* within the Hilti study. The lead user method lasted about nine months and cost US-$ 51,000. In comparison, conventional methods that lasted about 16 months and were twice as expensive (Herstatt/von Hippel 1992). An additional report by Meehan/Baschera (2002) stated that Hilti's customer focused programme Champion 3C continued to focus on lead users.

The major empirical evidence that the lead user method outperforms traditional market intelligence approaches for NPD was contributed by Lilien et al. (2002). The empirical research *interviewed divisional managers* in the multi-technology enterprise 3M and was conducted to analyse (1) contemporaneously funded lead user and non-lead user projects and (2) major product line ideas that were generated by lead users and non-lead users in the period from 1950 to 2000. The period for lead user based major new product lines (MNPL) could be traced back to 1997. Overall, 3M applied the lead user method in seven projects that are often used as reference projects for successful lead user integration (e.g. von Hippel et al. 1999). Five out of seven lead user projects were further funded and compared with 42 projects that were developed with alternative methods.

Table 18 Comparison of Innovation Projects Related to the Value of the Idea
(Reference: Illustration in Reference to Lilien et al. 2002, p. 1051)

Relevant Aspects	Lead User Projects (n=5)	Alternative Projects (n=42)
Novelty compared with competition	9.6 (out of 10)**	6.8 (out of 10)
Originality/newness of customer needs addressed	8.3 (out of 10)*	5.3 (out of 10)
% Market share in year 5	68%**	33%
Estimated sales in year 5 (million)	US-$ 146***	US-$ 18
Potential for entire product family	10.0 (out of 10)**	7.5 (out of 10)
Strategic importance	9.6 (out of 10)*	7.3 (out of 10)
Significance: *p<0.1, **p<0.05, ***p<0.01		

Table 18 provides an aggregated overview. The comparison was done by an evaluation of the expected benefit and potential in the market. Lilien et al. (2002) relied on funded projects, since an internal evaluation of these projects had already been conducted. Both project groups had equal initial conditions like a similar setting on the level of the enterprise's staff (e.g. comparable job levels and experiences). Lead user projects were found to be *superior in form of novelty and financial success* based on estimated sales in year five after market release. Expected sales will be significantly higher by a factor of eight ($p<0.001$) than expected sales of ideas from alternative projects – with respect to the integration of a 25% deflator (see Lilien et al. 2002 for a short explanation).

All five funded lead user projects became MNPL (cf. Lilien et al. 2002). The longitudinal findings revealed 26 major product lines. Overall, 19 lines were developed between 1950 and 1996 and seven lines were developed from 1997 to 2000. Five of the latter ones were identified as lead user developed MNPL. Those possess significantly ($p<0.05$) higher sales volumes in year five according to the revenue projections than non-lead user developed product lines. Further, 3M was able to *speed up their innovation process* significantly ($p<0.01$) in contrast to the employed non-lead user methods before 1997. However, the estimated average revenue forecast of MNPLs in year five is nearly double. This emphasises the potential of lead user contributions. Besides, the analysis showed similar estimations to fit organisational manufacturing capabilities. In sum, lead user integration improves the performance of NPD projects in the case of 3M.

Nishikawa et al. (2013) made similar observations in a company named Muji. The empirical analyses in Muji led to significant differences ($p<0.1$) in unit sales after three years, significantly ($p<0.1$) higher revenues of user-generated products with significantly ($p<0.1$) *higher gross margins* in the field of health and beauty products. This could be confirmed by findings in the furniture department of Muji. The gross margin tends to be four times higher in average (Nishikawa et al. 2013).

Further studies (see e.g. Al-Zu'bi/Tsinopoulos 2012, Carbonell et al. 2012, Tsinopoulos/Al-Zu'bi 2012, and Al-Zu'bi/Tsinopoulos 2013) highlighted *positive performance implications* in reason of using the lead user method (see also table 8 that fosters the motivation to incorporate lead users). Among them are positive aspects of increased productivity, decreased production costs, and innovativeness.

2.4.3. Relevance in Practice

Sections 2.4.1 and 2.4.2 point to successful applications and recommendations for lead user integration from a companies' perspective. Besides, revealed strengths and weaknesses of the method lead to the conclusion that the lead user method should be fostered as a standard tool for NPD in practice.

Lehnen et al. (2014) examined 255 business magazines in the German speaking area to point out the practical relevance and the *relationship between scientific contributions and their adoption in practice*. The 1st article in a German magazine was published in 1991 and started a trend of lead user-related articles that lasted up to 2009. About 50% of all articles were published between 2005 and 2009. The most frequent magazines were absatzwirtschaft, VDI Nachrichten, and Harvard Business Manager. About 59% of all articles just named the lead user method and only 41% of the identified articles described the method in detail. Overall, more than 40 detailed use cases were identified from the field of sporting goods, white goods, and special applications like the Tip Ex-case and the development of radical new coffee filters. Prominent larger firms that employed the lead user method were 3M, Hilti, and Johnson & Johnson aside with SMEs like Weidmüller Interface and Coppenrath und Wiese. The application of the lead user method was not only restricted to product development and covered service development, too (cf. Henne 2010 for an application example at the company DATEV). Additional 200 applications were mentioned in the literature.

Several scientific studies expressed concern about the *usage of the lead user method at an organisational level*. Lichtenthaler (2004) discovered three different approaches to coordinate the technology intelligence process. He identified lead user analysis as a *major method to design the information coordination task for technology intelligence* within the field of telecommunications equipment, automotive, and machinery industry. This task tries to manage the information gathering process to include ideas from external and internal sources. This approach exists next to structural and hybrid coordination. Lichtenthaler further argues that the importance of lead users in these markets is correlated with the relationship between the current technological progress and the current state of market development. Radical innovations in these markets seem to be mainly driven by regulations and thus lead user analysis does not play a notable role.

This is especially the case in the application field of pharmaceuticals (cf. Lichtenthaler 2004) although the method is known in this field (see e.g. Sänn et al. 2014). In sum, the usage of the method is dependent on the application field.

The need for radical innovations and its effective way to develop was further examined by Cooper/Dreher (2010). 160 companies were surveyed to rate various VoC methods according to their contribution to ideation. Based on the results, the usage of lead users was identified as "popular and effective" (Cooper/Dreher 2010, p. 42) with an average rating of 6.3 on a scale of 1 to 10. The method was extensively used in 23% of all interviewed companies. It is to mention that lead user analysis was interpreted as a VoC method (see Cooper/Edgett 2008).

Creusen et al. (2013) found similar high usage rates in Dutch consumer goods enterprises in general. Methods like (expert) interviews, focus groups, and questionnaire surveys were more prominent in the early phases of the innovation process – the early fuzzy front end. Interestingly, the knowledge of methods and their application in the context of NPD is described as being "limited" (Creusen et al., 2013, p. 82). The lead user method did not play a major role. "Maybe small companies find the [lead user] method too time intensive or do not possess the necessary knowledge about this relatively new method" (Creusen et al. 2013, p. 93). The authors mention that this is in contrast to previous studies.

Rese et al. (2015) studied VoC methods in the German automotive industry and highlighted the relevance of the lead user method for ideation and development with internal and external sources. Early findings revealed that among the interviewed companies (n=108) easy to use VoC methods, like questionnaire-based surveys, idea workshops or focus groups, are mostly used. Lead user integration was ranked fifth by its use frequency. This is consistent with the results of the multi-industry study by Barczak et al. (2009). Rese et al. (2015) included mainly respondents from automotive suppliers (56.5%) and from engineering service providers (17.6%) that lead to major feedback from SMEs (74.0%). The lead user method seems to be suitable for larger companies. If the method is known then it is also used and if it is frequently used then the evaluation of the method is significantly (p<0.01) positive, e.g. in quantity and quality of ideas of external sources (e.g. suppliers).

This is the case for larger companies. The usage of the lead user method is significantly correlated ($p<0.1$, $r=0.37$) with the *company size* (Rese et al. 2015).

Literature also provides evidence that *negative aspects of the lead user method* will become significantly ($p<0.05$) less problematic with an increasing usage frequency (cf. Al-Zu'bi/Tsinopoulos 2012 and Rese et al. 2015). Table 19 summarises the employed innovation methods in the fuzzy front end of NPD.

Table 19 Selected Empirical Studies of Methods in New Product Development

(Reference: Illustration in Reference to Rese et al. 2015)

Application Field	Employed Innovation Methods in the Fuzzy Front End	Reference Study
26 case studies from 3 fields (100% B2B)	Benchmarking studies (1), Portfolios (2), Flexible expert interviews (3), Scenario analyses (4), Patent frequency analyses (5), Quantitative conference analyses (6), Expert panels (7), Experience curves (8), Publication frequency analyses (9), Technology roadmaps (10), *Lead user analyses (11)*	Lichten-thaler (2004)
416 firms (59.1% B2B)	Beta testing (1), Customer site visits (2), Voice-of-the-customer (3), Alpha testing (4), *Lead users (5)*, Concept tests (6), Focus groups (7), Gamma testing (8), Ethnography (9)	Barczak et al. (2009)
160 U.S. firms (67.8% B2B)	Customer visit teams (1), Focus groups (2), *Lead user analysis (3)*, Customer advisory board (4), Customer brainstorming (5), Customer helps design product (5), Ethnography (7), Community of enthusiasts (8)	Cooper/ Dreher (2010)
88 Dutch firms (100% B2C)	Interview (1,1), Focus group (2,2), Complaint analysis (3,-), Segmentation (e.g. by demographic data) (4,-), Images/mood boards (5,-) ,Questionnaire survey (6,3), Brainstorming (-,4), Internet communities (7,-), Creating typical consumers (8,-), *Lead user analysis (9,6)*, Projective tech. (10,-), Observational research (11,5), Co-design (-,7), Conjoint analysis (-,8), User design (e.g. product configuration) (14,9), Scenario techniques (-,10)	Creusen et al. (2013)
453 firms (56.4% B2B)	Voice-of-the-customer (1), Customer site visit (2), Beta testing (3), *Lead users (4)*, Test markets (5), Alpha testing (6), Concept tests (7), Ethnography (8), Focus groups (9), Gamma testing (10)	Markham/ Lee (2013)
108 firms (86.1% B2B)	Idea Workshops (1), Netnography (2), Car Clinic (3), Focus Groups (4), *Lead User (5)*, Toolkits (6), Questionnaire Survey (7), Idea Competitions (8), Expert Interviews (9)	Rese et al. (2015)
In parentheses: ranking of usage frequency if method is known and applied; If two values are given: the 1st describes the ranking in the early FFE, the 2nd the ranking in the late FFE.		

The ranking for Lichtenthaler's (2004) study was for example derived from the intensity scale (often used – 3 points, sometimes used – 2 points, rarely used – 1 point, not used – 0 points) by the given data set for all three industrial application fields. This procedure was similarly used in analysing further studies, if there was no direct ranking provided. However, the presented methods for technology intelligence address different time horizons. This needs to be taken into consideration when interpreting the provided data.

Eisenberg (2011) also highlighted the importance of the lead user method in practice. The method faced a "... surge of popularity across companies and business schools in the late 1990s and early 2000s" (Eisenberg 2011, p. 50). Well-known articles like von Hippel et al. (1999) explained the application of the lead user method at 3M and emphasised positive and sustainable results.

Eisenberg (2011) summarised these studies and found a *variety of development goals*. Applications in 3M's Medical products division and in filtration focused on a new strategic direction. Cooling equipment focused on new markets and new applications. Packaging products division focused on new products and services. Biomaterials and commercial graphics division aimed at new technology platforms.

An analysis of 3M's web site shows that the development and market introduction of MNPLs were quite frequent in 2000 and 2001 with three new market releases per year. The following years seem to be characterised by an *innovation gap* of five years starting in 2006 and lasting until 2011. This confirms findings of Eisenberg (2011) that even 3M faced drawbacks with the lead user method. One explanation for the plunge assumes that introduced MNPLs failed their commercialisation and went off from the market (from the web site). This would be in contrast to the findings in previous studies (see Lilien et al. 2002). The analysis of the website revealed that 14 MNPLs were launched from the year 2000 to 2014 and are still present on the market. Among them are products like a digital signage system (that was sold to LSI Industries) and a molecular detection system to address business-to-business surroundings as well as a paint defender system and sports medicine equipment (sold under the brand of ACE) for consumers.

In contrast, the documented application at Hilti (see Herstatt/von Hippel 1992) was in fact a major success and marked the starting point for the *long-lasting new product line* of channel systems. This line was introduced to the market in 1992. Although no reliable data is published for this MNPL, the net income increased by approximately 47% from 1992 to 1993 and kept growing to double by 1997.

However, own experience from workshops and interviews support doubts that the method became a standard tool in SME's in Germany (see chapter 6 for details).

2.5. Derived Challenges

The lead user method faces challenges that can be derived from its methodological application, from psychological findings in literature, and from external restrictions.

The methodological application covers the challenge of a *reliable trend evaluation*, which is the basement for a correct lead user classification and identification. Von Hippel (1986) already mentioned reliable trend identification as an important aspect to perform a successful innovation task. Aside with this challenge is the significant *impact of the analyst* who is in charge to evaluate trends and potential lead users. The analyst faces the problem of *translating lead user needs* to future product attributes, but may not be able to understand the value of new ideas in reason of psychological restrictions. In contrast, lead users may show *individual needs that ordinary customers might not have in the future*. Then, the integration of lead users will result in a niche product development without fitting a broader market segment (see Magnusson 2009 and Mahr/Lievens 2012 for a discussion in reference to a suggested solution by Urban/von Hippel 1988 to predict adoption).

Research in the field of *classification of lead users* provided various approaches that may cause a misleading classification (see subchapters 2.2.4 and 2.3.3). This depends on the application field and deals with an unknown threshold to define the leading edge status or lead userness. Further, LES and lead userness may exclude promising contributions by lead users and may be *biased by self-evaluation* as could be observed by Belz/Baumbach (2010). However, this depends on the application focus – whether to gather contributions in quantity or quality.

The *identification of lead users* offers multiple approaches. The parallel procedure via screening provides a time advantage, but *inefficiency* is the consequence (see von Hippel et al. 2009 and Stockstrom et al. 2012 for empirical findings). This mainly depends on the size of the population. Today, decreased costs by using a web-based survey may favour screening over pyramiding. In contrast, due to the parallel search it is not possible to incorporate *learning effects* between the surveys (remarked by von Hippel et al. 2009). This is accompanied by incomplete information about the population, the lack of involvement of analogue markets that favours the *local search bias*, and a general unwillingness to participate in surveys (for example mentioned by Baier/Sänn 2013). Furthermore, the advantage of pyramiding in efficiency is significantly dependent on the respondent's interest in the market (r=0.44, p<0.05) and this has to be taken into consideration when dealing with FMCGs and low consumer involvement (cf. Kroeber-Riel et al. 2009).

This supports basic *psychological findings* like the *functional fixedness* (cf. Luchins 1942) of the project team, the *NIH* (cf. Antons/Piller 2014), and the local search bias (cf. von Hippel 1994). These findings do not necessarily affect the lead user itself but the project team. Especially, a team with highly specialised technological expertise faces the NIH (see e.g. Lichtenthaler/Ernst 2006, Kathoefer/Leker 2012, Sänn et al. 2013, and Antons/Piller 2014) and functional fixedness (see subchapter 2.2.2). Further psychological effects occur by neglecting internal creativity and outsourcing important innovation tasks to lead users. Thus, employees may stop to support this strategy, although their support matters (see e.g. Meehan/Baschera 2002) and needs to be strengthened by the *necessary commitment* of the senior management. However, the absence of commitment and supportive characteristics promotes rejecting behaviour and leads to attitudes like 'innovation without me' (see Wendelken et al. 2014). Another psychological aspect is the manner of collaboration within the lead user workshop. Literature points to the fact that lead users may dominate the workshop. This hinders collaboration and affects the necessary discussion among all participants (see e.g. Olson/Bakke 2001, 2004).

Several challenges that influence to application of the lead user method point to *external restrictions in resources*, e.g. a limited budget or time constraints. The empirical study of Lilien et al. (2002) revealed the fact that from a perspective of personnel expenses a lead user project is three times (p<0.05) more *expensive* than alternative development methods (154 person days vs. 60 person days; see Lilien et al. 2002, p. 1052). The method faces further difficulties in the question of *IP rights* that is triggered by the required transparency and by competitive aspects. Relying on embedded lead users may solve the aspect (see Schweisfurt/Raasch 2015). However, the *internal knowledge* about the methodological application may not be given in general. Further, managers are advised to *balance between radical innovation that favours novelty aspects and incremental innovation that favours usability aspects.* Within the surrounding of SMEs, managers might favour usability aspects to lower risks for the financial resources over novelty. The main goal is to stimulate an enhanced (accelerated) *diffusion* to cover a broad market (see Hinsch et al. 2014, de Jong et al. 2014 and von Hippel/DeMonaco 2013).

Table 20 summarises influencing aspects. In sum, these are valuable hints that will influence *uncertainties at the senior management's level* (see 2.4.1) and thus hinder the lead user method to become a standard tool in SMEs (see e.g. Olson/Bakke 2001) and in surroundings with restricted resources in general.

Table 20 Selected Challenges of the Lead User Method in Practice

Selected Challenges of the Lead User Method in Practice	Reference Study
Lead users have a strong interest that their own needs will be fulfilled exclusively.	von Hippel (1986); Olson/Bakke (2001)
Lead users may not be able to estimate product requirements in advance to ordinary customers.	Urban/von Hippel (1988); Mahr/Lievens (2012)
Lead users are not open-minded.	Lüthje (2004)
Lead user needs may not reveal future needs of customers.	von Hippel (1986); Creusen et al. (2013)
Internal staff does not accept lead user contributions.	Lichtenthaler/Ernst (2006); Kathoefer/Leker (2012)
The evaluation of lead user contributions is not easy (e.g. in relation to the market potential).	Enkel et al. (2005)
Lead user contributions may not fit in the company's focus.	Lichtenthaler/Ernst (2006); Kathoefer/Leker (2012)
Lead user contributions may not address a broad market.	Olson/Bakke (2001); Rese et al. (2015)
Open methods bind resources that are not available in SMEs.	Lasagni (2012)
The trust to integrate lead users is not given.	Lichtenthaler/Ernst (2006); Kathoefer/Leker (2012)
The evaluation of market and technological trends is difficult (e.g. in relation to the long-term impact).	von Hippel (1986)
The lead user method lacks in a broad support in the enterprise.	Lichtenthaler/Ernst (2006); Kathoefer/Leker (2012)
The theoretical knowledge to implement the Lead User Method is not available (e.g. for identification).	Enkel et al. (2005)
IP rights and protective policies are from major concern.	Lilien et al. (2002); Hoyer et al. (2010)

This is in line with Chesbrough/Brunswicker (2014) who found similar aspects that hinder open innovation in practice. Protection of internal knowledge, the NIH, and managing effort are major boundaries. This verifies previous statements and complicates the task of innovation managers (CIO/CTO etc.) in general.

3. Consumer Preferences

3.1. Overview

Surveying consumer preferences in a direct way is necessary to understand needs of ordinary consumers and the addressed market. Self-explicated measurement, for example, represents a compositional method. Typically, respondents are asked to rate attributes and attribute levels that are important for future products. This is done in different tasks and does not necessarily illustrate a complete stimulus. The preference values for stimuli are generated by composing independent values – a compositional method. In contrast, conjoint analysis is one of the state of the art techniques to measure respondents' preferences in a decompositional method. Green and Rao introduced conjoint analysis to the field of marketing in 1971 based on the approach of Luce/Tukey (1964). Traditional conjoint analysis (TCA) measures preferences on complete stimuli – a set of attributes that reflect a product. The method has been improved and extended to several variants like Adaptive- and Choice-Based Conjoint Analysis (ACA and CBC). Hybrid methods make use of both evaluation tasks. Compositional components in hybrid methods evaluate the importance of attributes and their related levels to prepare the stimuli sets. Decompositional components perform the preference measurement based on complete stimuli. Overall, respondents (ordinary customers and lead users that take part in a survey) tend to favour easy methods, since their mental effort to provide reliable data is relatively limited. However, the required effort increases dramatically with an increasing number of attributes and attribute levels, e.g. in CoPS. This was identified to lead to non-reliable and invalid results and is a common challenge in NPD (see chapter 2.2). Chapter 3.2 describes the theoretical background of consumer preferences and preference measurement. It introduces fundamental aspects of preference modelling and provides a brief historical overview. Chapter 3.3 briefly describes selected methods for compositional, decompositional, and hybrid preference measurement. This is extended by a discussion of findings in literature for developing complex products and systems (CoPS). Chapter 3.4 introduces collaborative filtering that aims to predict preferences on a computational basis. The methodological basics, its classification and selected applications are shown. Chapter 3.5 summarises encountered challenges that build the basis for the later Preference-Driven Lead User Method.

3.2.   Preference Measurement

3.2.1. Conceptual Fundamentals

Enterprises need to focus on innovations that will provide benefits for the consumer and will be preferred in a *real buying decision* (see e.g. Sawhney et al. 2006). Especially, *incremental product innovations* and *product line optimisations* rely on selected adjustments that are made to market-available products to address additional consumer needs or an extended target group. Incremental innovations rely on familiar products and technologies – or possibilities – to enhance a consumer's benefit (see e.g. Reichwald/Piller 2009). This can be seen as a basis to extend a current product line or to design niche products that address only specific needs. Such incremental innovations are, for example, additional colours, design changes to the bodywork, and additional equipment, e.g. in the automotive industry (see in reference Neibecker/Kohler 2009). Today, about 61% of development projects are from an incremental type (in reference to Cooper/Dreher 2010). Interestingly, this increased by 80% between the 1990s and the year 2010.

A consumer's benefit from preferring a combination of product properties over another is expressed and measured by means of preference values. *Preferences* describe the degree of favouring a certain product (the stimulus) over an alternative product on an individual level at a certain time (see e.g. Böcker 1986, p. 556). Thus, preferences describe a relation between two or more objects in terms of quantified differences of a consumer's attitude towards all objects and "... is needed to design new products or adopt existing goods to the customers' needs at reasonable costs" (Helm et al. 2008, p. 243). *Part-worths* indicate the preferences to a given variant of a product property. This is used to calculate the preference to a various sets of products properties variants for a respondent.

The methodological background to gather consumer preferences is given by *preference measurement*. Preference measurement determines the overall benefit of a certain product attribute for the consumer and reveals the consumer's willingness to pay (see e.g. Eggers/Eggers 2011 for an example from the automotive industry). This aggregates product properties with their evaluation by the respondent and reflects their relevance for the later buying decision.

However, preference measurement fosters on *attributes* with multiple *attribute levels* to describe a given object – the product. An attribute reflects a product property and an attribute level defines a specific variation for this attribute, like the body type of an automobile that is available as a sedan, as a saloon, or as a convertible. These alternatives are expressed as attribute levels. In general, this can be applied as dichotomous, continuous, or qualitative descriptions.

A *stimulus* describes a specific combination of attributes with a certain attribute level each. This reflects a product description by its attributes and levels that will be presented to the respondent (see Baier/Brusch 2009 for a basic introduction to preference measurement and requirements to find/define practical attributes).

Economic sciences point to the importance of a well-defined research object and research mission to foster a successful preference measurement. One of the most important aspects is the *correct definition of attributes and their levels* with respect to practical and theoretical requirements. "No decision is more critical [...] than the one that must be made about which attributes to include [in preference measurement]" (Auty 1995, p. 197). Auty noticed that attributes need to be actually involved in a consumer's trade-off decision. Elsewise, preference measurement will lead to false results and will not reflect the real buying decision.

Thus, the decision of how to *combine attributes and attribute levels* is relevant. This decision can be categorised in three levels: (1) the user level, (2) the subject level, and (3) the model level. The (1) *user level* reflects the consumer's point of view towards product alternatives. It requires that all combinations of attribute levels are (a) actually possible, (b) independent, which means that there are no interdependencies, (c) complete, which means that the respondent will have no chance to interpret missing properties, and (d) actionable (see e.g. Johnson/Levin 1985, Shocker/Srinivasan 1974 for a detailed discussion). The (2) *subject level* reflects the consumer's decision-making process and its psychological constraints. This describes the (a) relevance of an attribute and (b) the limited quantitative capacity of a respondent. The (3) *model level* demands (a) compensatory attribute relations, (b) preference undependability, and (c) omitted non-acceptable attributes and attribute levels that are described as "totally unacceptable" (Srinivasan 1988, p. 297; see also and Weiber/Mühlhaus 2009 for a brief summary).

For example, Weiber/Mühlhaus (2009) draw an example to illustrate the relevance of airline safety and discuss this attribute as being choice-deterministic or not in reference to Green/Srinivasan (1978). The basic assumption is that a respondent would generally assume that airlines are safe to fly. Thus, an unsafe airline would never be chosen as being preferable and thus airline safety should not be used as an attribute to generate a stimulus.

A further important aspect is the limited capacity of the respondent. If the amount of attributes and attribute levels exceeds the respondent's capacity, then the respondent will be exhausted, their attention will be deteriorated, and they become distracted (Rao 2014 speaks in terms of *respondents' fatigue*). It is commonly accepted that this is the case when more than seven attributes are in question. Several studies focused on this issue (see e.g. Sänn/Baier 2012, Bensch et al. 2012 in reference to DeSarbo et al. 2005, and Chen et al. 2009) and agreed that this leads to lower validity (see also subchapter 2.2.2).

Aspects of *reliability* and *validity* express the quality of preference measurement (see e.g. Brusch 2005 for a summary of previous findings). Reliability describes the formal accuracy of the measurement. A reliable measurement will lead to an equal result in a further measurement under an equal system definition – the results can be reproduced. For example, the speedometer in a car shows the same amount of mph under equal conditions (see Brusch 2005 for a description using a scale). Validity describes the objective accuracy of the measurement and the extent to which the measurement corresponds with the real world. For example, the car's built-in speedometer is showing reliable but wrong data. Typically, this requires another measurement with a different method – like using GPS data in this case. In marketing and innovation management, a valid preference measurement reflects the real world buying decision (see Leigh et al. 1984 for a brief description and an early comparison of preference measurement methods in relation to validity). Predictive validity expresses the accuracy of the predicted decision in comparison to a real decision that can be observed. This can be measured by using holdout cards. Spearman correlation and the FCHR may serve as indicators to show the correlation with the real world. Today, preference measurement is the most important technique to stimulate incremental innovations. This is placed in the phase of *concept generation* (see figure 1) in NPD.

3.2.2. Preference Modelling

In the 1970s market intelligence was concerned about the prediction of consumer behaviour and was in need to understand their attitudes. Underlying psychological models were employed to describe the cognitive structure of consumers and to predict their buying decision. Typically, products consist of multiple attributes with several opportunities to design them. This reflects the basic thought of a multi-attributive model. It evaluates a product with its multiple attributes and further possible combinations of attribute levels. In general, the assumption is that the attribute-wise combination of attribute levels will result in a benefit for the consumer that influences the buying decision (e.g. Wilkie/Pessemier 1973).

Literature in the field of marketing argues that traditional preference models resulted from the fundamental *expectancy-value approach* that was adopted from psychology and is intended to explain a respondent's acting. This action is motivated by two indicators (see Mazis et al. 1975 for a basic introduction and a comparison with further models):

(1) the strength of the expectancy that the act (e.g. a buying decision) will be followed by a consequence – for example an expected improvement of one's own image while buying a specific automobile brand, and

(2) the value of that consequence to the person – the benefit from improving one's own image.

The *Fishbein model* is also seen as an early development to determine attitudes of respondents to given objects (cf. Fishbein 1967). The attitude towards an object is expressed by an individual belief that a behaviour will result in an outcome. This attitude towards an object is multiplied with the individual value importance.

Aside with Fishbein's application in behavioural science, *Rosenberg (1956)* presented a similar model that was linked to political concerns in the 1950s. However, the attitude towards a political action is expressed by the sum of the perceived instrumentality of a value state and the individual value importance.

Rosenberg aimed at political possibilities that depend on other people's reactions. Fishbein's model was formulated without this influence. For example, the question if a brand is capable of improving one's social status may lead to different results in reference to both intentions. Mazis et al. (1975) criticised that. Both models expect perceived instrumentality and belief strengths. Both aspects "... would be high whenever the individual [the consumer] strongly believe that the value is achieved regardless of how satisfactory the value itself is perceived to be" (Mazis et al. 1975, p. 41). However, expectancy multiplied with importance was the common approach to determine preferences (Mazis et al. 1975).

The *adequacy-importance model* (Cohen et al. 1972) employs the importance of the attribute and the evaluation of the object with respect to the specific attribute. The adequacy-importance model was shown to be favourable over the expectancy-value approach in various experiments (see Mazis et al. 1975).

The *vector model* argues further that the Fishbein model "… does not discriminate between perceived possession and probability [of possession]", e.g. when evaluating an object to contribute to a specific outcome (see Ahtola 1975, p. 53 and Cohen et al. 1972 for further details). Ahtola (1975) describes this with an example on "Coca Cola is very carbonated" (p. 53) which would be operationalised to the attribute "carbonated" without any information about the amount of carbon dioxide in it. Thus, respondents may be confused, e.g. by misinterpreting the probability scale, and a demanding approach would be needed to cover multiple mediators. This missing mediator may also affect the evaluation of the importance. The vector model presents this demanding approach by incorporating a mediator (importance) to generate the preference model. The main advantage is the separation of "… strengths of the belief from the content of the belief" (Ahtola 1975, p. 58) to result in a specific outcome. This emphasises the dependency of the stimuli evaluation from the favoured attribute levels. Athola (1975) has shown that his approach outperforms the Fishbein model. The vector model fosters on a linear correlation between increasing attribute levels and the resulting benefit. An example of the vector models is fuel consumption of a car. The less fuel is consumed by a car, the more preferable is the car. This is not true in every case and would foster on a wrong preference model, e.g. when performance is taken into consideration or when this vector model is applied for preferences at a concert hall with maximum sound level.

The *ideal point model* assumes that an ideal attribute level exists. Any deviation from the ideal point leads to a decreased evaluation of the specific attribute of a stimulus. This is applicable when using continuous attribute levels and is for example used in the *Trommsdorff model* (see Bichler/Trommsdorf 2009). For example, any deviation from an ideal mixture of pharmaceuticals will lead to non-preferable stimuli. A similar example can be drawn with the taste of toothpaste.

The *part-worth model* is the common model in practice (see Green/Srinivasan 1978 and Green/Srinivasan 1990). Overall, preference models are used to express the relative importance of attribute levels on an individual basis and to calculate part-worths in accordance with the underlying utility function (see Wright/Kriewall 1980 for a discussion). The part-worth model builds the basis for e.g. conjoint analysis.

A summary of early preference models is given by Green/Srinivasan (1978). An actual summary with a focus on conjoint analysis is presented by Rao (2014)

3.2.3. Utility Modelling

The *utility function* (also utility model) describes how preference values for multiple attributes are aggregated. The utility expresses attraction for a given outcome (see Edwards 1954) and is the basis for economical behaviour. Eggers/Eggers (2011) provided an example of the part-worth model for preference measurement in the automotive industry and used a *linear utility function*. This function describes that the part-worth of an attribute can be simply added to the part-worth of another attribute to generate the overall (complete) utility of a stimulus. Alternative functions are either logarithmic or exponential and depend on the prediction/assumption of the customers' relative importance of an attribute to assume the individual function.

Additive utility functions belong to the field of *compensatory models*. Compensatory models assume that a (positive) part-worth for a specific attribute level is balanced by a (negative) part-worth of other attribute levels from the same attribute. Part-worths compensate each other (see e.g. Elrod et al. 2004). Further, compensatory models assume that there is no unacceptable attribute and thus a consumer would decide to buy a product even if a specific attribute that is present in every stimuli would be rejected. For example, a respondent would consider to choose a car despite its (unwanted) high fuel consumption as long as this car is manufactured by a desired brand, provides acceptable performance, and offers an appropriate level of comfort (all attributes with their specific levels).

Non-compensatory models assume that no substitution of attributes and attribute levels is given. This leads to a (1) *conjunctive* and (2) *disjunctive* approach to understand consumers' utility model within a market.

The (1) conjunctive approach assumes that the buying decision is driven by specific attributes and thus the elimination of all stimuli that cannot achieve an expected (possibly minimum) level from the respondent's perspective will be the result. This is "… rejection of any object that fails to meet a minimum criterion on an attribute" (Elrod et al. 2004, p. 1) and describes that a decision may be determined by a single attribute. For example, a respondent would reject any car that does not provide a top-speed of 150 mph. The related (2) disjunctive approach assumes that the respondent would consider any stimulus as favourable as long as it provides a desired attribute level. This "…results in acceptance of an object that surpasses a very high standard on at least one attribute, irrespective of its values on the other attributes" (Elrod et al. 2004, p. 4). For example, a respondent would accept any car that does provide a top-speed of 150 mph.

Thus, the accurate modelling of consumer's utility model is the most important point in preference measurement. This includes the determination of key attributes (e.g. fuel consumption that is influenced by social and economic trends) to assume conjunctive and disjunctive approaches, but needs to frame the stimuli with an appropriate amount of attributes (see also in reference to the analyst's influence to lead user projects in subchapter 2.2.2). It is therefore the key aspect to accurately model how respondents will achieve their *maximal utility*.

Compensatory models like conjoint analysis by Green/Rao (1971) and the self-explicated method by Srinivasan (1988) prevailed as dominant methods for preference measurement and represent a compositional and a decompositional approach each (see e.g. Hensel-Börner 2000, Chrzan/Golovashkina 2006, Sattler 2006, Park et al. 2008, Eckert/Schaaf 2009, Netzer/Srinivasan 2011, Scholz et al. 2010, and Meißner et al. 2011 for methodological adaptions and empirical comparisons with simple products and CoPS).

3.3. Methodological Applications

3.3.1. Compositional Methods

Compositional methods use a direct evaluation (free eliciting) of product attributes and attribute levels (see e.g. Akaah/Korgaonkar 1983 for a comparison of different methods and Johnson 1987, pp. 259-260 for an exemplary description). Traditional approaches of compositional preference elicitation concentrated on attributes, only (see Jain et al. 1979). Modern approaches design a compositional preference measurement by (1) evaluating all attribute levels for a specific attribute at once, (2) ranking the importance of the attributes, and (3) aggregating results to calculate the overall utility. The most prominent compositional methods are the Analytic Hierarchy Process and the Self-Explicated Measurement (see Saaty 1986 for AHP and Srinivasan 1988 for SEM).

The *AHP* (see Wind/Saaty 1980 for its application in marketing) "[...] structures any complex, multiperson, multicriterion, and multiperiod problem hierarchically" (Wind/Saaty 1980, p. 641) under the application of trade-off decisions. A detailed reflection of AHP's (dis-)advantages is given by Saaty (1994a) next to a detailed description of its application (Saaty 1994b).

Typically, *SEM* (Srinivasan 1988) measures attributes and attribute levels separately in mostly two stages and can cover a preparing stage 0 that describes the elimination of unacceptable attribute levels (see the discussion by Hensel-Börner 2000, p. 19). Stage (1) describes the evaluation of acceptable attribute levels and stage (2) refers to the evaluation of the attribute importance. However, a basic classification of SEM refers to stage one and stage two. Stage two can be omitted and defines SEM as being a *weighted or unweighted* method. Unweighted SE methods employ an *exclusive evaluation of attribute levels* (stage 1). Methods that make use of stage one and stage two are described as weighted, since the evaluation of attribute levels is weighted by the evaluation of the related attributes. Srinivasan (1988) further highlights the use of an anchor that is supposed to define a critical attribute from which the importance rating of other attributes is obtained.

Today, several variants exist to employ SEM in practice as it is given by e.g. Eckert/Schaaf (2009, p. 45). The authors summarised and categorised multiple evaluation techniques to operationalise the measurement in both stages in reference to previous literature. They further revealed undiscovered combinations of both stages with respect to the use of specific measurement approaches.

Scholz et al. (2010) combined SEM and AHP to form the so-called *Paired Comparison–Based Preference Measurement* (PCPM). Meißner et al. (2011) compare Adaptive Self-Explication Method (ASE), Adaptive Conjoint Analysis (ACA), and PCPM. It was shown that ASE is to be prefer for complex tasks.

Netzer/Srinivasan (2011) published the *Adaptive Self-Explication Method* (ASE) that aimed at CoPS in terms of dealing with a relatively large number of attributes and attribute levels. This weighted SEM illustrates a web-based method and "...breaks down the attribute importance question into a ranking of the attributes followed by a sequence of constant-sum paired comparison questions between two attributes at a time" (Netzer/Srinivasan 2011, p. 141). The major benefit is the resulting reduction of pairwise comparison tasks that is supposed to lower a respondent's effort and the related distortion (see chapter 3.2.1). The missing results of the more comprehensive traditional comparisons are derived from the attribute level rank order. "Specifically, respondents are asked to evaluate how valuable the improvement from the least to the most preferred level of each attribute is to them" (Netzer/Srinivasan 2011, p. 141). Thus, the importance of an attribute itself can be derived and an additive preference model applied.

Schlereth et al. (2014) provide a detailed look to SEM with trade-off decisions. This reference can serve as an overview for the state of the art in self-explicated preference measurement. The authors introduce the *Pre-Sorted Adaptive Self-Explicated Approach* (PASEA). PASEA adopts ASE with modifications on the ranking task of attributes. "That is, instead of immediately asking respondents to perform the ranking [of attributes] in stage 2, we ask them 1st to rate the improvement from least to most preferred level of this attribute on a 7-point scale. These ratings serve to pre-sort the attributes [...]" (Schlereth et al. 2014, p. 188).

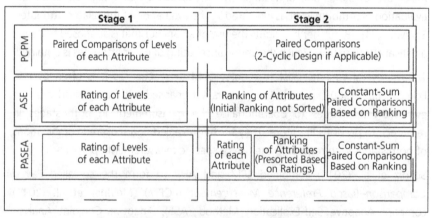

Figure 8 Methodological Summary of PCPM ASE PASEA

(Reference: Illustration in reference to Scholz et al. 2010, Netzer/Srinivasan 2011, and Schlereth et al. 2014)

Figure 8 provides a methodological overview to PCPM, ASE, and PASEA. It illustrates the employed measurement for the 1st stage (preferability of attribute level) and the measurement for the 2nd stage (importance of attributes). ASE and PASEA differentiate the 2nd stage by e.g. employing an initial ranking that is followed by a paired comparison using constant sum rating.

3.3.2. Decompositional Methods

Decompositional methods employ preference measurement based on complete products – the stimuli set. This means that the product is defined by its attributes with a specific attribute level and represents a composed product (see e.g. Wilkie/Pessemier 1973). Typically, decompositional preference measurement uses multiple stimuli to perform trade-off decisions by the respondent.

If the stimuli set is generated with all possible attributes of a product, then the measurement follows the *full-profile approach* (see e.g. Green/Srinivasan 1978). The opposing *trade-off* approach is characterised by using two attributes (see Johnson 1974 for an exemplary study on trade-off analysis). Overall, the respondent is asked to compare two or more presented stimuli and to choose one of them. The measured total utility values of the stimuli are decomposed to part-worths for the attributes and attribute levels. Practically, one assumes a *compensatory and additive decision-making process with a linear utility function* to evaluate the presented stimuli (see e.g. Green/Rao 1971).

Summarised, "... any decompositional method that estimates the structure of a consumer's preferences [...]" (Green/Srinivasan 1978, p. 104) is described as *conjoint analysis*. Luce/Tukey (1964) introduced the concept of a simultaneous conjoint measurement to the field of behavioural science that originally corresponded to the measurement of the effects of two functions that imply two different classes of variables. The main difference to established (compositional) preference measurement was that a concatenation of two functions, e.g. x(a) and x(b) is not the sum of x(a+b) in every case. Conjoint analysis introduces a function to define the influence of x(a) and x(b) to their summation, which is for example relevant for physical observations. By its definition it "[...] is concerned with the joint effect of two or more independent variables on the ordering of a dependent variable" (Green/Rao 1971, p. 355), which adopts the Luce/Tukey (1964) method. Green/Rao (1971) brought conjoint analysis to the field of marketing in the 1970s.

In *Traditional Conjoint Analysis* (TCA) respondents are asked to rank concept cards, which represent the stimuli. TCA uses a full-profile approach, employs a linear additive utility model, and starts with an initial selection of attributes and attribute levels. The necessary input may be derived from creativity methods like morphological analysis and attribute listing.

In sum, when n represents all attributes and m represents their attribute levels then there would be a maximum of m^n *possible stimuli*. This maximum amount of resulting stimuli needs to be reduced to an appropriate amount to build the stimuli set. This can be done by using an *orthogonal design* (see Addelman 1962) that requires independency between the given attributes (see chapter 3.2.1) and generates a reduced design.

The resulting ranking from the evaluation task represents the dependent variable as an order of stimuli cards and illustrates the order of preferable product concepts by the respondent. The individual utilities of the independent variables (part-worths) are calculated typically using *Ordinary Least Squares regression*. The overall importance of the attributes is generated by the maximum differences of the resulting part-worths per attribute. This spans a utility vector that is interpreted in relation to the overall deviation of all attributes (see Green/Wind 1975 for an example applied in package design). The last step describes the aggregation of all individualised results. Brusch (2005, pp. 25-28) summarises the progress.

Today, multiple versions of conjoint analysis exist. The most important decompositional version is the *Choice-Based Conjoint Analysis* (CBC, see Louviere/ Woodworth 1983 and McFadden 1986). One major breakthrough of this version is that respondents do not rate or rank stimuli in a direct way, but perform a *discrete choice* with a non-buy option that fosters on a full-profile approach. Another relevant adaption is the usage of a *Maximum-Likelihood estimation* to calculate the utility parameter (see e.g. Schmittlein/Mahajan 1982 for the usage of the estimation in innovation diffusion), among others.

Selka (2013) has shown that *CBC is the primarily applied version* of conjoint analysis in practice. However, the author showed that external validity (FCHR) significantly ($p<0.01$) dropped in decade from 2003 to 2011. This may be related to an *increasing complexity* of the objects (products) and an *increasing distraction*, e.g. by Facebook (see e.g. Bensch et al. 2012 and Rao 2014). An overview of the conjoint analysis' evolution in the 1980s and the 1990s is given by Green et al. (2001) and further in Rao (2014, pp. 345-360).

3.3.3. Hybrid Methods

Hybrid methods for preference measurement consist of a decompositional and a compositional approach. Compositional methods – namely the self-explicated method – are used to determine the desirability of each attribute level separately and to measure the related attribute importance. The decompositional – conjoint – model is used to perform a limited amount of trade-offs for a reduced set of full-profile evaluations (see Green 1984 for a description of hybrid methods).

The *Adaptive-Conjoint Analysis* (ACA, see Johnson 1987, Johnson 1991, Green et al. 1991, Toubia et al. 2007, and Johnson/Orme 2007) is one exemplary hybrid method. It is structured in five steps with: (1) calibration of inacceptable attribute levels, (2) preference ranking, (3) relevance rating, (4) pairwise comparison, and (5) the likelihood measurement of the buying decision.

Park et al. (2008) presented another adaption with a focus on complex products – *Web-Based Upgrading* (WBUM). The respondent starts the preference measurement with a minimal configuration of a product and upgrades this product stepwise with pre-defined and interdependent attributes and levels. This further analyses the respondents' WTP and can be seen as a product configurator/toolkit.

In general, the complexity of products has increased in recent years and thus preference measurement became a very extensive and demanding task for the respondent (Netzer et al. 2008, p. 243). This is especially the case when large amounts of attributes and attribute levels are employed to describe the stimuli (see Rao 2014). The overall intention of hybrid methods is therefore to increase the validity of preference measurement. This is expected to be achieved by *lower data collection time and effort* (cf. Green 1984 and Green/Krieger 1996) and is aside with a reduction of methodological complexity of the measurement task.

Previous studies showed implications for *practicality reasons* like ease of use (Akaah/Korgaonkar 1983) and cognitive stress for the respondent (see Sattler/Hensel-Börner 2007, p. 71). Also Scholz et al. (2010) and Meissner et al. (2011) expanded the focus in comparing methods for preference measurement by aspects of a real buying situation, ease of use for the respondent, and the overall survey length among others. In general, multiple studies point to favour self-explicated measurement over conjoint analysis in complex product evaluation tasks because of these reasons, although conjoint analysis reflects a real buying decision (see e.g. Akaah/Korgaonkar 1983, p. 188 and Sattler/Hensel-Börner 2007, p. 71). The study of Meißner et al. (2011) further shows that respondents favoured ASE over ACA, e.g. because of its usability.

Academic literature argues about favouring conjoint analysis over self-explicated preference measurement and vice versa in reason of *reliability* and *validity* (see e.g. Hartmann/Sattler 2004). Early studies indicated that self-explicated measurements provide better reliability (see e.g. Heeler et al. 1979, Leigh et al. 1984, Green et al. 1993, and Hensel-Börner 2000, pp.45-48). This points to the assumption that conjoint analysis (decompositional and hybrid) seems to be more capable for simple products and self-explicated measurement (compositional and hybrid) should be used within complex product development (see e.g. Netzer/Srinivasan 2011 and Meißner et al. 2011). Recent studies show contradictory results in predictive validity and reliability (see Sattler/Hensel-Börner 2007, pp. 72-73 for a summary). Pullman et al. (1999) and Oppewal/Klabbers (2003) highlighted conjoint analysis because of its increased validity. Scholz et al. (2010), Meissner et al. (2011), Netzer/Srinivasan (2011) added recent findings for complex products (see table 21).

Table 21 Selected Studies on Preference Measurement in Complexity

(Reference: Illustration in Reference to Sattler/Hensel-Börner 2007, Scholz et al. 2010, Netzer/Srinivasan 2011, and Meissner et al. 2011)

Application Field	Results of Comparison	Reference Study
Cell Phones (14 Attributes)	PCPM provides sign. higher FCHR than ACA.	Scholz et al. (2010)
Coffee (8 Attributes)	SEM partly better predictive validity than computerized CA.	Hensel-Börner/Sattler (1999)
Digital Cameras (12 Attributes)	ASE with significant higher hit rate than ACA.	Netzer/Srinivasan (2011)
Electronic Blenders (10 Attributes)	SEM approach with better reliability than TCA.	Heeler et al. (1979)
Laptop Computers (14 Attributes)	ASE with significant higher hit rate than ACA.	Netzer/Srinivasan (2011)
Refrigerators (9 Attributes)	No significant difference, ACA performs better in case of predictive validity.	Huber et al. (1993)
Summer Vacation (10 Attributes)	PCPM provides sign. higher FCHR than ACA.	Scholz et al. (2010)
Theme Parks (15 Attributes)	No significant differences in predictive validity between ASE and ACA. ASE with improved applicability (reality, survey length, ease of use).	Meissner et al. (2011)

Besides actual findings, dealing with a large amount of attributes and attribute levels remain a problem (see Rao 2014, pp. 218-220 for an evaluation of selected methodological versions and Verlegh et al. 2002 for range implications and the number of levels effect). Thus, no generalizable answer can be provided on whether conjoint analysis is to be favoured over self-explicated measurement.

Overall, preference measurement offers a variety of methodological approaches, but the *methodological advantages are dependent on the product category, complexity, and the practicability* of the method for the application field. Findings also showed the importance to distinguish between high- and low-involved respondents and the need to take the survey's target group into account as well. Sattler/Hensel-Börner (2007, p. 69) aggregated advantages of self-explicated measurement and conjoint analysis that shall be taken into consideration before the decision is made by the innovation manager, such as *reducing a social response bias, detection of non-linearity in the utility function, and data collection expenses.*

Traditional self-explicated preference measurement (SEM) will be used in the prototype of the Preference-Driven Lead User Method. This decision is fostered on practicability reasons (speed, cost, and ease of use) for respondents and for later implementation. Also, preference measurement is applied to cover core product attributes and thus one can expect sufficient results (see Srinivasan/Park 1997).

3.3.4. Relevance in Science and Practice

A conducted *database analysis* in EBSCOhost, Web of Science, and ScienceDirect with the keywords "conjoint analysis" and "conjoint measurement" (in reference to Brusch 2005, p. 11; see also subchapter 2.3.2) found 4,362 publications by September 30, 2014. Duplicates are not filtered out from these results. Conjoint analysis-related publications reached their annual peak with about 144 articles in 2011 in EBSCOhost, 74 articles in ScienceDirect and 72 articles in the Web of Science. In 2014 EBSCOhost lists only 55 articles, ScienceDirect 79 articles, and the Web of Science 31 articles. Figure 9 summarises the results of the database analysis. The results from the analysis in subchapter 2.3.2 serve as a reference. This may indicate that conjoint analysis faces a declining adoption in academia (see in reference to Rogers 1962, Bass 1969, Bass et al. 1994, and Bass 2004).

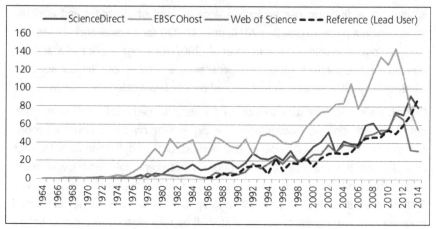

Figure 9 Annual Academic Publications from 1964 to 09/2014 (CA)

A historic overview of its *importance in practice* is given by e.g. Cattin/Wittink (1982), Wittink/Cattin (1989), Baier (1999), Voeth (1999), Hartmann/Sattler (2002), Brusch (2005, p. 23), and Teichert/Shehu (2010). Wittink et al. (1994) found that 44% of the surveyed market intelligence companies used conjoint analysis. The market intelligence company TNS Infratest performed an average of 200 annual CBC studies between 2002 and 2011 (see Selka/Baier 2014). Selka (2013, p. 72) confirms the major application in incremental product optimisation (52.41%) but also points to challenges in validity.

3.4.   Prediction of Preference Data

3.4.1. Definition and Classification

The treatment of missing values in surveys is a constant topic in marketing science in general and preference measurement in particular (see e.g. Little/Rubin 2014 and Decker/Wagner 2008 for an introduction to the basic problem, a historical overview of related methods, and a discussion on their applicability). In this thesis, the focus is on the problem of *item-non-response*. The necessary prediction of missing data for a respondent becomes interesting when, for example, a consumer's preference of a stimulus cannot be gathered by survey techniques, like in a dynamic questionnaire, or if the *preference value is simply missing*. This may be the case if data were lost during data exchange, intentionally deleted, or an evaluation was skipped by the customer. This occurs by randomness, *by design*, or by non-response on purpose and is caused by multiple reasons such as the exhaustion of a respondent.

Decker/Wagner (2008) highlight the relevance in practice and illustrate that especially surveys with many (>200) items are affected by this. Previous research like Kim/Curry (1977) mentioned the loss of statistical accuracy in reason of missing values. Actual literature indicate that a *10% amount of missing data is considered as adequate* (see Schnell et al. 1999, Tsikriktsis 2005 for a review of 103 articles in operations management dealing with missing values).

Literature differentiates multiple types of causes for missing values. *Missing at random* (MAR) types are unrelated to the respondents' true status on that item, but related to another item. In other words, no dependency between the missing value and the item is given, but are caused by another item. An often-cited example (e.g. cited by Tsikriktsis 2005) deals with answers to the questions of educational level and income. MAR is described as being present, when (1) the probability of an answer to a dependent item (the income) between multiple respondent groups (multiple educational levels – independent item) is different and when (2) within a group (a specific educational level) the answer to the dependent item (income) is unrelated to its level (amount of income) (Tsikriktsis 2005, p. 55). For example, the higher the education the more likely is an answer to the question of income, but the amount of income is independent from the probability of providing an answer.

Missing completely at random (MCAR) types of missing values are further characterised by the lack of observed relationships between the dependent (probability to answer income) and the independent item (educational level).

Missing completely at random within classes (MCARC) enables a separation of the respondents into classes of missing values. MAR and MCAR(C) types of missing values can simply be ignored by further analysis to determine the underlying pattern of missing values (see Decker/Wagner 2008, p. 60 for a description).

Not missing at random (NMAR) types of missing values show (significant) differences within the groups of the independent and dependent item. For example, the probability to answer the question of income is dependent from (1) the educational level and (2) the amount of income.

In practice, listwise or pairwise deletion – known as *case reduction* – of the data set is used to skip missing values next to an estimation – known as *value imputation.* Deletion and imputation were found to cause – in most cases – inconsistency and inefficiency (see Little 1988 and Decker/Wagner 2008 for a detailed discussion).

In general, Kim/Curry (1977) showed that a combination of a large sample size, a small proportion of missing information, and a basic randomness of the missing information is favourable to produce valid estimations on missing data in surveys. Moreover, a random independency between the variables in a survey and information about the respondents on a meta-level is suggested to perform valid estimation. Otherwise, missing information would lead to a comparison of different kinds of respondents and variables. Thus, multiple methods evolved to impute missing values in survey data. An example is mean imputation that is based on the mean value (see Göthlich 2009 for a summary of methods).

"Cold Deck" and *"Hot Deck" imputation* benefit to value imputation by making use of similarity and a nearest neighbour approach. Both approaches "…correspond to imputing means within subclasses of the sample" (Little 1982, p. 244) and are a convenient solution in practice. The fundamental approach assumes that contextual 'near' respondents provide also 'near' survey data.

This is contextually analogue to *collaborative filtering* that describes a set of computing algorithms used within a recommender system to filter information and makes use of an assumed relationship between respondents and between items (variables in a survey). The main purpose of this type of information retrieval is the identification of desired and undesired data from a set of information (see Resnick et al. 1994 for an early application).

Early thoughts on collaborative filtering tasks can be traced back to the case of Tapestry. Goldberg et al. (1992) argue that the underlying information selection can be modelled as collaborative work in which people help each other to *filter interesting and uninteresting documents* in a mailing system (in other words: in an information stream) by recording their annotations to the document. The Tapestry approach used known social relationships and similar annotations as background information for the information filtering task. An annotation can be an endorsement by a specific moderator of the mailing list as well as the knowledge (in reference to the social relationship) that selected individuals – who reply to a document – share a common interest. Thus, only documents that were sent to those specific persons are of interest for a user (see Goldberg et al. 1992 for a detailed description of Tapestry and further filter mechanisms). This is the 1st version of collaborative filtering and was required due to the desire for an effective spam filter to "manage" information overload.

Resnick et al. (1994) experienced similar circumstances and noticed major complaints about information overload and a low signal to noise ratio among all incoming information, like in the Usenet. The authors make use of the idea that people who agree in their evaluation of past articles tend to agree on future articles, too. The system *correlates the ratings* and *predicts how well users will like future articles* based on ratings of other users (see Resnick et al. 1994).

A set of information can also be presented as a stimulus or a product, like in the case of Amazon's recommendations. Aside is the *prediction of missing information* that can be employed to foreshadow later buying decisions. Literature on innovation management noted a possible applicability of collaborative evaluation within a community (see e.g. Hienerth/Lettl 2011 and Jensen et al. 2014).

However, collaborative filtering in this thesis is used from the computer science perspective. The basic algorithm creates a *recommendation list of top items* for user u_a with only a little trace of its preference values. The basic assumption is that users who agreed on the rating for some items typically also agree on the rating for other items (see e.g. Baltrunas et al. 2010 and Hahsler 2011, p. 3). This is similar to 'hot deck' imputation. Thus, collaborative filtering is used to *complete missing user ratings* based on given data from numerous other users for multiple items to perform a prediction of preferable product attributes.

In theory, this can be handled either for *single users or for joint consumption processes* that may be applicable in a business-to-business surrounding (as shown by Hennig-Thurau et al. 2012). The basic thoughts on such automated recommender systems implies that a *collective wisdom helps consumers to make better choices* than without such a decision support system (Hennig-Thurau et al. 2012, p. 90 in reference to Krishnan et al. 2008).

In general, collaborative filtering uses an underlying *user-to-item rating matrix* $R=(r_{jk})$ with a set of rating values r for all users u_j and for all items i_k. The missing ratings for user u_a (called the active user by Hahsler 2011) are predicted by calculating the similarity between each user. This creates a complete new row r_{ak}

in R that includes the predicted rating data. The basic requirement is that the active user u_a and other users u_j share an item i_k that was rated by both.

Model-based approaches "learn" (Hahsler 2011, p. 3) from the user database and calculate recommendations with a fraction of all users. Memory-based approaches (see e.g. Ghazarian/Nematbakhsh 2014) are using the complete user database to create recommendations. The calculation promises results that are more accurate when the number of users increase (see Resnick et al. 1994 on scale effects).

3.4.2. Item- and User-based Collaborative Filtering

The basic assumption of item-based collaborative filtering is that *users will prefer items that are similar to the items they already like* (Hahsler 2011). This approach is suitable for products and services that are consumed repetitively like FMCGs or movies, music, and books (see Hennig-Thurau et al. 2012 for an example).

In contrast, user-based collaborative filtering focusses on the users' preferences (see e.g. Herlocker et al. 2004). The basic assumption of user-based collaborative filtering is that *users with similar preferences will also rate items in a similar way* (see Hahsler 2011). This similarity is based on preferences for the rated items in the matrix and is suitable for heterogeneous products. Users with similar ratings define the neighbourhood for the active user u_a. The similarity is calculated by e.g. using the Pearson correlation and requires at least one rated item that is shared with another user u_m. Another approach to calculate the similarity is using the Cosine distance (see for example Su/Khoshgoftaar 2009 for an overview).

The initial step is to calculate the correlation or distance to identify neighbours. The *neighbourhood* is defined either by the k-nearest users to u_a or by setting a threshold value for the similarity in general (Hahsler 2011). This produces a matrix (u,u') that reveals information about the similarity of all users in the set.

When a suitable neighbourhood is found, the prediction for the missing ratings r_{ak} is performed, for example by *averaging the ratings* of the neighbourhood. A weighted average would respect the fact that some users are more similar to user u_a than others. Further, a normalisation of the ratings is recommended to remove the rating bias when users tend to rate low or high for all items (see Hahsler 2011). Table 22 introduces a brief example to compose a neighbourhood of user u_a. The table provides information of the ratings for user u_1 to u_6 and u_a. Items i_1 to i_8 were rated, but not every user rated them accurately.

Table 22 Rating Matrix for the User-Based CF Example

(Reference: Illustration in Reference to Hahsler 2011)

Item/ User	i_1	i_2	i_3	i_4	i_5	i_6	i_7	i_8
u_1	?	4.0	4.0	2.0	1.0	2.0	?	?
u_2	3.0	?	?	?	5.0	1.0	?	?
u_3	3.0	?	?	3.0	2.0	2.0	?	3.0
u_4	4.0	?	?	2.0	1.0	1.0	2.0	4.0
u_5	1.0	1.0	?	?	?	?	?	1.0
u_6	?	1.0	?	?	1.0	1.0	?	1.0
u_a	?	?	4.0	3.0	?	1.0	?	5.0

The given user u_a rated only items i_3, i_4, i_6 and i_8. In an initial step, the similarity of each user with u_a is calculated using Pearson correlation or Cosine distance. Table 23 provides the results for both. In fact, a Pearson correlation might fail when a user constantly rates the same value for all items or both users share only one common item. This is known as the problem of a cold start.

Table 23 Similarity Estimation with Pearson Correlation and Cosine Distance

User/User	u_1	u_2	u_3	u_4	u_5	u_6
u_a (Pearson)	0.756	-	0.866	0.982	-	-
u_a (Cosine)	0.961	-	0.937	0.996	-	0.832

It is to mention that user u_3 becomes more similar under an assumption of Pearson correlation. Users u_2, u_5, and u_6 remain outside the neighbourhood in both cases. This may change if a threshold for Pearson correlation and Cosine distance is used instead of a fixed number (k=3). The ratings for the missing items of user u_a are calculated by the average of the neighbours' ratings for item i_j.

Table 24 Calculated Ratings r_a based on the Neighbourhood

(Reference: Illustration in Reference to Hahsler 2011)

Item/ User	i_1	i_2	i_3	i_4	i_5	i_6	i_7	i_8
u_a	?	?	4.0	3.0	?	1.0	?	5.0
r_a	3.5	4.0	-	-	1.3	-	2.0	-

Table 24 provides the missing ratings r_a for user u_a. A further Top-3 list would recommend items i_8, i_2, i_3 to the active user u_a.

3.4.3. Applications in Practice

Collaborative filtering algorithms are a major research topic (Shi et al. 2014 provide an exemplary survey of the state of the art) but are also widely used in practice, like in the domain of *online retailing* (see e.g. Stüber 2011 for a summary). Often referenced application scenarios in this field describe hedonic goods like movies, books, video games, and wine (see for example Hennig-Thurau et al. 2012 and Cacheda et al. 2011).

The online-marketplace *amazon.com* gives an example. Amazon was founded in 1994 and opened its online-store in 1995. Amazon.com already served 1.5 million customers and gained a revenue of $147.8 million in the year 1997, which was one of the most important years for the company with a growth in sales by 838% and a growth of 738% in customer accounts. The inventories in these days were counted to over 200,000 book titles. Today, amazon.com is one of the largest online-marketplaces (next to alibaba.com) covering multiple product categories like the traditional and electronic book market, apparel and shoes, movies, music, apps, and live video streaming among other areas. This led to a net revenue from products and services of about $74 billion in 2014 (for a detailed look on Amazon's historical data and actual financial statements see Amazon 2014a). Amazon.com describes the basic functionality of its recommendation algorithm with the statement *"We examine the items you've purchased, items you've told us you own, items you've rated, and items you've told us you like. Based on those interests, we make recommendations."* (Amazon 2014b).

Linden et al. (2003) provided a more detailed description of Amazon's own developed item-to-item collaborative filtering implementation. Amazon's success led to multiple limitations of traditional collaborative filtering approaches. Among these limitations are *real-time functionality, limited information for new customers and volatile customer data*. An own development was needed to overcome previously mentioned obstacles and to realise an own "... algorithm's online computation [that] scales independently of the number of customers and number of items in the product catalogue" (Linden et al. 2003, p. 76). The collaborative filtering implementation "... matches each of the user's purchased and rated items to similar items, then combines those similar items into a recommendation list. To determine the most-similar match for a given item, the algorithm builds a similar-items table by finding items that consumers tend to purchase together" (Linden et al. 2003, p. 78). An iterative process follows and is given in pseudo code below.

FOR each item in product catalogue I_1 {

    FOR each customer C who purchased I_1 {

        FOR each item I_2 purchased by customer C {

        RECORD that a customer purchased I_1 and I_2 {}}}

    FOR each item I_2 {COMPUTE the similarity between I_1 and I_2 {}};

The algorithm identifies each consumer with a positive purchase decision for the 1st product in the catalogue. Then the algorithm identifies every other product the consumer has bought and marks (records/annotates) those pairs of items ($I_{1,2}$). At that point, the similarity is calculated for each identified product that the consumer bought next to item I_1. The Cosine distance is used to compute the similarity between items $I_{1,2}$. These steps are repeated for every product in the catalogue (see Linden et al. 2003 for the pseudo code).

The original patent of Amazon's recommender implementation is filed under US patent number 6,266,649 (see patent information from 2001 and 2012 for further details). Interestingly, the patent contains multiple approaches to generate various recommendations. Besides, Amazon's recommendation algorithm is a well-classified secret that possesses *significant influence on its business success*.

Today, Amazon and other e-commerce companies are able to use further detailed data to frame their recommendations. Among them are attributes like *items viewed* and *click stream data*. Moreover, in-memory data management provides future benefits for real-time capability and faster computing of the (former offline) similar-items table that favours user-based collaborative filtering. This enables, for example, a fast and reliable *online search* based on collaborative filtering (see e.g. Linden's 2005 patent for Microsoft Corp. under US 2005102282 A).

Academic research has shown implementations of collaborative filtering like in the Jester Online Joke Database (Jester 2014). Jester uses a collaborative filtering algorithm called *Eigentaste* (see Goldberg et al. 2001) to recommend jokes to an active user based on her or his own ratings of previous jokes. Jester starts with a *calibration phase* to gather ratings on pre-selected jokes. The basic user interface provides the possibility to rate between "less funny" and "more funny" which describes the interval between -10 and +10. This phase covers five jokes and is *equal for every user*. The active user is asked to rate further presented jokes. Recommendations of jokes that the user may like are based on the initial ratings and are becoming more precise with every new rating of a joke. Today, Jester assigns users to (1) Eigentaste or (2) Eigentaste 5.0 (see Nathanson et al. 2007) randomly to compare both algorithms. The application Donation Dashboard is another example of Eigentaste that was patented by UC Berkeley in 2003.

In the following, a randomly selected 100 users and 100 items sample of the Jester database is used to illustrate user-based collaborative filtering in progress. Table 25 presents the user-item matrix for a fraction of the database. It shows jokes 1 to 100 and 100 randomly selected users. The user-item matrix is the relevant input for further collaborative filtering. The R package recommenderlab (see Hahsler 2014) was used for calculations. The matrix covers 7,862 ratings. The focus is on User 2056 (in *italics*) who rated joke i_5, i_7, i_8, and i_{10}.

Table 25 Original User-to-Item Matrix for the Jester Sample

Item/User	i_1	i_2	i_3	i_4	i_5	i_6	i_7	i_8	i_9	i_{10}	...	i_{100}
920	6.75	8.59	-0.15	-0.68	0.87	3.50	6.46	-0.68	-2.23	6.07	...	-0.39
1695	1.70	-7.48	-	-	0.05	-	-6.31	-9.90	-	-2.28	...	-
2056	*-*	*-*	*-*	*-*	*-0.49*	*-*	*1.26*	*0.10*	*-*	*7.52*	...	*-*
2192	-1.26	0.73	-4.42	-5.05	-1.80	5.29	-4.42	-5.58	4.08	-4.90	...	2.38
2252	7.09	-6.07	-6.50	-4.27	7.23	5.68	-0.44	3.35	1.89	6.17	...	-
2256	-3.50	3.83	-3.88	0.63	-0.10	-1.02	-8.40	-1.99	0.24	2.14	...	4.17
2282	5.83	0.53	4.90	-0.73	-4.56	3.30	-0.29	3.11	-1.21	8.69	...	-4.61
2644	-	4.42	-	-	-0.78	1.07	4.13	3.06	-	-3.45	...	-
2872	-	-6.55	-	-	3.54	-	-1.75	-9.42	-	-	...	-
2974	2.52	5.15	1.80	3.98	2.62	3.50	4.08	-0.87	1.70	5.58	...	3.30
...	...	...	...	...	...	...	...	...	...	...	...	...
24865	8.83	-6.31	8.83	4.17	-0.29	-6.31	-4.76	5.73	8.83	8.83	...	8.83

It is the main objective to predict user's (u_a) ratings for the non-rated items. This starts with a calculation of the similarity matrix (see subchapter 3.4.2). In this example, the Cosine distance is used since the Pearson correlation was already introduced in the last subchapter. The computed preference evaluations for the previous missing values are printed bold in table 26. As a result, user 2056 receives joke i_1 as the top recommendation.

Table 26 Updated User-to-Item Matrix for the Jester Sample

Item/ User	i_1	i_2	i_3	i_4	i_5	i_6	i_7	i_8	i_9	i_{10}	...	i_{100}
920	6.75	8.59	-0.15	-0.68	0.87	3.50	6.46	-0.68	-2.23	6.07	...	-0.39
1695	1.70	-7.48	**-2.98**	**-5.58**	0.05	**0.41**	-6.31	-9.90	**-3.79**	-2.28	...	**-0.73**
2056	**4.21**	**3.83**	**3.53**	**2.30**	-0.49	**3.17**	1.26	0.10	**3.86**	7.52	...	**3.32**
2192	-1.26	0.73	-4.42	-5.05	-1.80	5.29	-4.42	-5.58	4.08	-4.90	...	2.38
2252	7.09	-6.07	-6.50	-4.27	7.23	5.68	-0.44	3.35	1.89	6.17	...	**0.85**
2256	-3.50	3.83	-3.88	0.63	-0.10	-1.02	-8.40	-1.99	0.24	2.14	...	4.17
2282	5.83	0.53	4.90	-0.73	-4.56	3.30	-0.29	3.11	-1.21	8.69	...	-4.61
2644	**2.18**	4.42	**1.91**	**1.00**	-0.78	1.07	4.13	3.06	**1.15**	-3.45	...	**1.93**
2872	**-4.48**	*-6.55*	**-4.98**	**-6.01**	*3.54*	**-4.69**	*-1.75*	*-9.42*	**-4.51**	**-4.85**	...	**-5.39**
2974	2.52	5.15	1.80	3.98	2.62	3.50	4.08	-0.87	1.70	5.58	...	3.30
...	...	...	...	...	...	...	...	...	...	...	...	...
24865	8.83	-6.31	8.83	4.17	-0.29	-6.31	-4.76	5.73	8.83	8.83	...	8.83

Overall, collaborative filtering offers opportunities to deal with missing values and to predict preference data. The prediction of the acceptance of new (lead) user contributions within a community can be assumed to be a promising avenue to evaluate the commercial success that helps to find the needle in the haystack.

3.5. Derived Challenges

Preference measurement is an important technique to determine consumers' needs and preferences for incremental innovations. Still, only 15% of all corporate efforts are dedicated to successful products that fulfil profit expectations (as further confirmed by Cooper/Kleinschmidt 2000 and highlighted by Schneider/Hall 2011). One of the reasons can be found within the very difficult prediction and evaluation of consumer's needs. Thus, the prediction of buying decisions remains vague. A discussion on challenges of preference measurement in general covers respondents, methodological restrictions, and the surroundings.

The open and user innovation community (see chapter 2) argues that *ordinary consumers are not aware of their own needs* and are *unable to express their needs in a definite and reliable way* leading to an information problem (as could be observed by Jeppesen 2005). Further, ordinary consumers cannot *think beyond their own use experience* and tend to be limited by their *functional fixedness*. This may lead to a failure of formal procedures to understand customers' preferences (as seen by Eisenhardt/Martin 2000 and Narver et al. 2004) in reason of *missing reliable and valid data*. For example, the lack of knowledge and experience of ordinary customers influences the reliability of VoC techniques. This is because of the fact that consumers possess *latent needs* and change their mind frequently (Slater/Narver 1998, Mahr/Lievens 2012). Today, consumers are confronted with a great product variety and thus choosing the right product seems to be a challenging task. They are in need of support to foster their decision making process (as mentioned for example by Sarwar et al. 2000).

The methodological perspective on challenges of preference measurement covers preference and utility modelling. Both tasks are highly dependent on the *experience of the analyst* and need to cover an average consumer in a specific market. However, each consumer possesses an own utility function. Thus, it may not be applicable to assume a compensatory, conjunctive or disjunctive utility model in general. Literature further highlights that the *complexity of a product* influences the validity and reliability of preference measurement (see for example Helm et al. 2008 for an observation in the field of mountain biking). CoPS lead to complex measurement tasks and this is accompanied by the number of levels effect (Steenkamp/Wittink 1994) and *distortion* (Rao 2014).

Moreover, former utility-driven buying decisions are influenced by hedonic product attributes as Dhar/Wertenbroch (2000) showed. For example, the former utility-based buying decision to choose a new automobile is now accompanied by hedonic attributes, like a sporty (unique) design that is hard to shape into standardised questionings. Thus, "... [a] whole new approach is needed [...] that accurately respond[s] to users' needs" (von Hippel 2001, p.247).

Recommender systems are capable of improving this situation. However, it is to notice that collaborative filtering operates under circumstances with "... only a *small fraction of ratings* [that] are known and for many cells in [matrix] R the values are missing" (Hahsler 2011, p. 3). Thus, predictions become less accurate with descending input data.

In practice, leading e-commerce websites have many item pairs with no common customers and this makes *computation inefficient*. Only a small fraction of products is even bought by one customer in the Amazon example (Linden et al. 2003). This is known as a problem of *sparsity* (see e.g. Sarwar et al. 2000 for the basic problem and Huang et al. 2004 for an approach to alleviate it). Sparsity describes the circumstance that there is only little information about positive buying decision available. For example, within a large online marketplace, "... even active customers may have purchased well under 1% of the products [in stock]" (Sarwar et al. 2000, p. 161). Thus, the identification of similar users in relation to user u_a becomes a challenging task and this results in inefficiency and inaccuracy.

Subsequently, nearest neighbourhood approaches are *computational expensive* and suffer from a problem of *scalability* (see Hahsler 2011, Sarwar 2000, and Resnick et al. 1994) and *real-time processing*. Further boundaries of collaborative filtering algorithms are the so called *"lemming effect"* (see for example Stüber 2011, p. 30 with further references) as well as problems with new objects, new users and first recommendations after initialising the recommender system – known as the problem of a *cold start* (Schein et al. 2002).

Sarwar et al. (2000) shows the problem of *synonymy*. Initial data analysis highlights the problem of *over-recommending* popular users (Krzywicki et al. 2015). Further future challenges are given by Shi et al. (2014). Su/Khoshgoftaar (2009) and Cacheda et al. (2011) summarise (dis)advantages of memory- and model-based collaborative filtering.

Summarised, collaborative filtering within an offline environment like in the prototype of the Preference-Driven Lead User Method is able to overcome most restrictions, since it is used in a non-real-time system (omitting scalability as a challenge and cold start problems) with a calibration phase that addresses the problem of sparsity. Further, non-e-commerce applications are able to use a *retrospective view* and aim to fill missing ratings that obsoletes the lemming effect. Even the synonymy problem is omitted since the project team moderate new items in advance. Overall, collaborative filtering seems to be a beneficial module to model the prototype. Further, ordinary customers are able to evaluate basic contributions (cf. Heiskanen et al. 2007). Kornish/Ulrich (2014) provided empirical evidence that *early-stage evaluations are unbiased* and that a crowd of ordinary consumers is able to perform as a sufficient evaluator to predict market outcomes. Preference measurement can help to distinguish between involved and non-involved respondents. This may help to exclude non-reliable respondents from further analysis and to improve the handling of missing values by collaborative filtering.

4. The Preference-Driven Lead User Method

4.1. Overview

The Preference-Driven Lead User Method focuses on challenges that are derived from the traditional lead user method. The major methodological adaption is formed by addressing the basic research question *"Can the lead user method and preference measurement be combined to result in an integrated method for new product development and outperform the standard sequential development?"*. It proposes a combination of both methods to generate contemporary preference data for internal and external (lead) user contributions. This is enabled by incorporating collaborative filtering before the lead user workshop is hosted. The resulting method consists of five phases that cover preparation, trend identification, lead user identification, preference measurement and the lead user workshop. In contrast to a sequential progress of ideation and evaluation (see figure 1), the proposed method combines both methodological steps from NPD and handles them jointly within one method. The prototype performs an evaluation of internal and external contributions. Thus, the method is expected to be able to select feasible contributions to be further processed within the workshop – regardless if the contributions were made by a lead user or an ordinary customer. Overall, the Preference-Driven Lead User Method is expected to result in an advanced product concept by terms of feasibility, market potential and novelty. It is assumed that market potential can be increased. Novelty is expected to be decreased. Feasibility is described by (1) market acceptance and (2) fit to the available resources of a company. The derived knowledge allows an estimation of internal development costs in advance of the workshop. The workshop can further concentrate on favoured contributions. The development of product concepts is jointly performed with internal developers, lead users, and available basic preference data. Besides, it is intended to address a senior management's uncertainty by providing preference data and to overcome the project team's functional fixedness as well as the NIH. The latter one may be achieved by acceptance values for internal contributions. The remainder of this chapter derives the concept from findings in literature in chapter 4.2. It provides detailed insights from the phases and notes the implementation (chapter 4.3). Chapter 4.4 provides an application example from mountain biking. Chapter 4.5 provides implications.

4.2. Conceptual Basis

4.2.1. Fundamental Challenges

The joint development with lead users in the fuzzy front end of the NPD process can provide benefits and shows that user developments are able to overcome development restrictions in SMEs, as discussed in chapter 2. In summary, a joint development between a company and lead users can increase the productivity, decrease production costs, and generate positive influence to its innovation performance, e.g. in terms of novelty and product advantages (see table 8). However, the project team and its environment influence the methodological application in several phases. Methodological aspects that are influenced by the quality of the project team and determine the output of the traditional lead user method are for example (see chapter 2):

(1)   the broad definition of the focus,
(2)   the evaluation and identification of recent trends,
(3)   the classification of lead users,
(4)   an identification of lead users,
(5)   the integration of analogous markets,
(6)   the evaluation of lead user contributions, and
(7)   a lack of knowledge about broader market segments.

Further aspects that influence the methodological application are the underlying psychological challenge of *functional fixedness* of the project team, the *local search bias*, and the *complexity of a problem*. Moreover, needs and contributions of lead users require an in-depth understanding of the problem space. Thus, the detailed *assessment of those needs and contributions* cannot be simply drawn by non-lead users' input only (in reference to Urban/von Hippel 1988, Jeppesen 2005, Enkel et al. 2005, Magnusson 2009, Mahr/Lievens 2012, Creusen et al. 2013, Rese et al. 2015, who pointed to a gap between lead users and non-lead users in their quality to contribute to joint NPD projects).

Basic challenges that go beyond previous argumentation are summarised by recent literature: "The limitations of the lead user approach are its high potential risk (e.g., [...] that the wrong individuals or firms are identified as lead users). Furthermore, [...] lead users' concepts can fall victim to the *NIH*" (Hienerth et al. 2013, p. 3).

A lack of fundamental *knowledge* about how to apply the lead user method in an appropriate way was also noted by Creusen et al. (2013). Further aspects cover the leading role of lead users within the *workshop* next to the overall *influence of the analyst* and its psychological constraints itself. Especially SMEs that deal with limited resources (see for example Lin 2003 for a general discussion and Rammer et al. 2012 for an analysis within the German economy) face the latter aspect in reason of their awareness for *technological, commercialisation, time, and financial risks*, which will immediately affect their liquidity and future existence on the market and, thus, innovation projects remain high-risk endeavours (see Cooper/Kleinschmidt 1987, Kleinschmidt/Cooper 1991, and Cooper/Kleinschmidt 2000) for SMEs. These fundamental aspects lead to *uncertainty* (as summarised by Olausson/Berggren 2010) and a *lack of trust* in the application of the method.

In sum, these aspects lead to the important question that is answered by Urban/von Hippel (1988): "But *what if lead users like the product and non-lead users do not?* In this case, there are two possibilities: (1) The *concept is too novel* to be appreciated by non-lead users – but will later be preferred by them when their needs evolve to resemble those of today's lead users; (2) the *concept appeals only to lead users and will never be appreciated* by non-lead users even after they 'evolve'" (cf. Urban/von Hippel 1988, p. 580). Both cases are serious challenges and pose major risks for future existence, especially for SMEs.

In the 1st case, diffusion is highly questionable since the point in time of the adoption of the concept by non-lead users remains unknown. In the end, this leads to a failed or *long-term diffusion* process (what is in contrast to expectations from addressing lead users to speed-up the diffusion process, see Schreier et al. 2007). In the 2nd case, this circumstance is clearly a *major risk for future existence* since only a niche market would be served and this may results in a development failure.

Hinsch et al. (2014) point to the relevance of *diffusion promoters and the relevance of early adopters* to support the diffusion of user innovations. Remarkably, they highlighted the relationship between the diffusion of a new technique, in their cases described as medical techniques, to stimulate subsequent user product innovations. Thus, the *influence of non-lead users* to the diffusion process of user innovation is relevant for a successful innovation.

De Jong et al. (2014) give a more factual observation. Only 19% of all innovations in a sample of 176 innovations that were developed for personal use by customers in Finland led to a successful diffusion and did not lead to a market failure. Here, the authors describe *market failure as the need of increasing effort by the innovator to promote his user innovation to stimulate an adoption.* In contrast, the adoption by early adopters etc. would lead to a successful diffusion and less effort by the innovator is required to stimulate this. In sum, non-lead users may be a notable source to enhance the diffusion (Hienerth/Lettl 2011) and are able to provide *reliable judgement* over radical developments (cf. Heiskanen et al. 2011) in the form of basic expressions. Another basic aspect is given for example by the buying centre (e.g. given by Webster/Wind 1972) and frames an *unreliable decision making process in markets for industrial goods* that include multi-person structures. Today, this multi-person decision is an actual topic in the industrial decision-making processes (e.g. Bigler/Drenth 2013) and thus a product needs to cover multiple perspectives in this group.

The usage of *preference measurement* can be taken into consideration to decrease the before mentioned general uncertainty. Methodological aspects that determine reliability and validity of preference measurement are for example (see chapter 3):

(1)   the choice and number of attributes,
(2)   the product's complexity,
(3)   the awareness of needs,
(4)   the reliable expression of needs,
(5)   the assumed utility function,
(6)   the ratio of missing data, and
(7)   the experience of the analyst.

However, preference measurement may serve to ve*rify the output of the method before the final workshop* is hosted. An additional clustering of the derived preference values can then be incorporated to the workshop to design a new product concept with lead users to fit specific needs of the market. Following this, the (lead user) toolbox can be refined for usage in SMEs (see Lasagni 2012 for a discussion and criticism, and Creusen et al. 2013 for a demand to address the *very specific needs of SMEs*). This postulates the *Preference-Driven Lead User Method.*

4.2.2. Methodological Adaption

Observations from previous chapters point to multiple adjustments that can be made in practice like adjusting classification, identification, and the workshop setting. The fundamental idea to shift preference measurement to the fuzzy front end covering the ideation and evaluation task connects the process of 'ideation and evaluation' with 'concept generation' and spans lead users and preference measurement to both stages (see in reference to figure 1) as a joint method.

This fundamental idea is derived from the fact that *a similar evaluative structure of lead users and non-lead users foreshadow a positive adaption* by the target market (cf. Urban/von Hippel 1988). This is emphasised by findings that ordinary customers are able to provide basic evaluations of new product concepts (cf. Heiskanen et al. 2007) and raw conceptual ideas (Kornish/Ulrich 2014). Von Hippel/von Krogh (2013) speak in terms of need-solution pairs and proposed that the *discovery of a need is correlated with the discovery of a solution* and happens at the same time. Edl et al. (2014) observed that about 45% of new developed need-solutions pairs described this scenario. Thus, a need becomes real and cognisant when solutions for a prior undiscovered problem are presented. This is also known as the term of "latent needs" (see e.g. Kroeber-Riel et al. 2009).

This setting becomes interesting when ordinary consumers are asked to take part in preference measurement with *lead user contributions*. This combination of the lead user method and preference measurement can be done in two directions: (1) as a follow-up – sequential – approach and (2) as an implementation of preference measurement before the lead user workshop.

The *1st approach* describes the traditional lead user method with successive preference measurement (see for example Sänn/Baier 2012 and Sänn et al. 2013). This illustrates the basic process of NPD, analyses a generated concept to be further developed, and possesses no further *influence to the result of the lead user method*. If preference measurement indicates negative results, then the *innovation process needs to start all over again*. For example, it could be seen that the *derived concept from the lead user method will fail* within preference measurement, the costs for the lead user project were already paid, and resources were already spent. Besides, promising concepts may not be shifted to preference measurement due to a previous cancellation by the upper management (see Behrens/Ernst 2014 for insights into decision-making structures for go and strop decisions that are biased). This may be the case in a scenario with increasing uncertainty (as pointed out in chapters 2.4 and 2.5). Thus, a failed project will increase the general uncertainty and *hinder further investments of resources to further concept development* using the lead user method (see for example Olson/Bakke 2001 and Enkel et al. 2005).

The *2nd approach* incorporates *preference measurement* as a standalone phase within the lead user method (see table 28). It shifts the phase of "projecting lead user data onto the general market of interest" (von Hippel 1986, pp. 802-803) *in advance of the lead user workshop* – the Preference-Driven Lead User Method.

This new method requires an interconnection of the traditional lead user method and preference measurement with an utilisation of collaborative filtering (see chapter 3). It reflects a promising way to answer the main research question *"Can the lead user method and preference measurement be combined to result in an integrated method for new product development?"* and to overcome mentioned restrictions (see subchapter 4.2.1). Table 27 illustrates applied adaptations and compares the new method to the traditional lead user method (in reference to von Hippel 1986).

Table 27 Adaptions to the Traditional Lead User Method

	Phase 1	Phase 2	Phase 3	Phase 4	Phase 5
von Hippel (1986)	Identification of market- & technology trends	Identification of lead users	Analysis of lead user needs (& ideas)	Applying lead user data to target market	-
Preference-Driven Lead User Method	Preparation to launch the lead user project	Identification of trends and internal contributions	Exploration of the user community	Preference measurement	Lead user workshop

Hienerth/Riar (2014) confirmed the positive influence of a crowd for evaluation tasks. Both authors adopt Eric von Hippel's formulation that (trend) evaluation remains "something of an art" (von Hippel 1986, p. 798) and examine the use of crowds to perform the necessary task.

Behrens/Ernst (2014) noted that the evaluation by external consultants "[...] might be an effective approach to prevent managers from furthering commitment to a losing course of action" (p. 368) since the evaluation by the managers solely is influenced by a decision bias. Other literature speaks in terms of NIH to describe this observation (see Antons/Piller 2014).

Related research in this field points to the importance of this evaluation task. In other words, building a *micro-community* around (lead) user contributions may overcome common boundaries of an evaluation task and address management's uncertainties while leveraging to a *macro-community* (see Hienerth/Lettl 2011).

Cooper (2011) demands extensive to *p management commitments to product innovation* to stimulate successful innovations that might not be given under *uncertainties*. The necessity of this commitment was empirically proven and confirmed to stimulate a development of superior products (see e.g. Graner/Mißler-Behr 2013 who analysed this circumstance in 410 NPD projects).

4.2.3. Basic Concept

The 1st phase of the Preference-Driven Lead User Method describes the basic *project grounding*. The project grounding defines the setup of the project in terms of resource allocation and project planning tasks. This includes team building, framing of the project scope and definition of the master project plan. This phase was kept untouched and is applied in reference to Churchill et al. (2009).

The 2nd phase deals with the *identification of trends and internal contributions*. The identification of market- and technology trends includes trend screening and the definition of an attribute set for preference measurement of state of the art attributes. The identification of major trends follows the traditional method (see e.g. von Hippel 1986), but *the evaluation task is omitted here*.

The *development of internal contributions* is further included in this phase and builds the bridge to the survey design in the following phase. The project team is expected to *define standard attributes and attribute levels* for the product that is in the development focus (see chapter 3.2). Internal contributions from the project team serve *as starting points to foster the later survey* for enhanced contributions. Thus, it is suggested that the internal ideation is extended to include actual developers and later workshop participants, if applicable. Later, this input also serves as *motivational factors* for the community to participate in this firm-hosted ideation task (e.g. to address extrinsic and intrinsic motivational factors like expanding one's own ideation and stimulating learning; see subchapter 2.2.5).

The 3rd phase describes the *exploration of the user community*. The *lead user classification* can be done by multiple approaches which include using the traditional definition, focussing on an idea development or employing terms of lead userness or leading edge behaviour (see subchapter 2.2.3). Operationalised characteristics are derived from the innovation focus that was provided by the senior management. This specification influences the survey design.

The *survey design* shall reflect *motivational aspects* through graphical, textual, and structural design. It is also valuable to think about a gaming design but this leads to other scientific research streams. However, the design is also highly related to the favoured *method of preference measurement* (see subchapters 3.3.1 and 3.3.2) and depends on the amount of addressed trends, attributes, and their levels.

The *survey distribution* is fostered on the preferred method to perform the *identification of lead users* (see subchapter 2.3.4) that can be done by traditional means of the screening approach, by broadcasting, signaling, and pyramiding. The latter one would use this survey as an interview guide and employs preference measurement in a CAWI version (see Selka 2013 for an overview).

Elsewise, preference measurement can be separated from lead user identification and will be sent to the respondent after the informal interview. In fact, the main intention in this phase is either to achieve a *maximum market coverage* and a *maximum amount of external contributions* or to concentrate on qualitative links and efficiency using the pyramiding approach. If quantity is in focus, one would choose the screening approach, if applicable with the given resources. The distribution may rely on a motivational design and thus benefit from viral effects.

The 4th phase of the Preference-Driven Lead User Method describes the *survey and reference analysis*. The analysis insures *data quality*, covers the classification of lead users, and preference measurement for state of the art attributes as well as internal and external contributions in the form of acceptance values.

Overall, this is expected to prepare the workshop by *providing information* about:

(1)   the validity of applied preference measurement,
(2)   preference structures of lead users and non-lead users for SOTA attributes,
(3)   acceptance values for internal and external contributions,
(4)   acceptance and preference structures of lead users and non-lead users, and
(5)   specific preferences within a buying centre (optional).

Internal and external contributions are analysed using binary measurement. Toubia/Florès (2007) pointed to the circumstance that companies often face a *high amount of* ideas and need to evaluate them by preference measurement. The authors employed binary evaluation to perform idea screening. The reader may imagine a setting like in the IBM Innovation Jam with hundreds of thousands participants and tens of thousands contributions (see Bjelland/Wood 2008). This thought is lent to frame preference measurement for non-standardised contributions and thus binary measurement is appropriate to collect first experience with the prototype of the Preference-Driven Lead User Method.

This is also *caused by multiple reasons*, among them: (1) detailed preference measurement for broad contributions would lead to invalid results (see Heiskanen et al. 2007), (2) detailed preference data is expected to bias the later development process (e.g. because of a pseudo-accuracy), (3) complex preference measurement may lead to respondent's fatigue (in reference to Toubia/Florès 2007 and Rao 2014), and (4) it is expected that detailed rating data will be compromised since innovative contributions are continuously added to the survey. This may influence the respondent's cognitive rating process and lead to incomparable data.

However, literature shows that ordinary customers are able to understand basic concepts well enough to rate them either useful or not (see Heiskanen et al. 2007 for customers' conservativeness and understanding of radical innovation concepts), but they tend to *fail in the task to favour one concept over another* – as it is done in traditional preference measurement – without detailed background information or use experience. However, internal and external contributions are rated by every respondent with a binary coding and will be analysed in order to address the NIH by providing feedback to internal engineers and developers for their contributions.

Collaborative filtering is used to provide an evaluation of external contributions and uses the present binary values for internal contributions as calibration variables.

The 5th phase describes the traditional *lead user workshop*. Derived from the 4th phase, the project team is able to *pre-assess own and external contributions* in terms of feasibility, production, and resource allocation to estimate the effort for later NPD. Additional *similar evaluation structures* of lead users and non-lead users in relation to these contributions allows to predict a later adoption and to become aware of heterogeneity or homogeneity of the market.

In reason of this process in the 5th phase, the innovation manager receives information from preference measurement and idea pre-assessment to *prepare the workshop*. The *structured concept development* at the workshop is intended to solve detailed challenges to aggregate preferred contributions and to generate future work packages for production. This workshop might be performed using external lead users or exclusively rely on internal engineers and developers. The lead user workshop can be organised in reference to Churchill et al. (2009) to build a concept based on the gathered internal and external contributions.

Afterwards, the senior management will decide whether to *continue or to stop the project's further development*. This is the interface to continue with further stages of traditional NPD like testing and production ramp-up (see figure 1).

The Preference-Driven Lead User Method is illustrated with its phases in comparison to a modern lead user method in table 28. The comp*arison of both methods* is based on Eisenberg's (2011) paper "Lead-User Research for Breakthrough Innovation". Eisenberg's experience from the application of the lead user method at 3M, Bell Atlantic (today Verizon), and Gillette (today Proctor & Gamble) is merged in her article with recent findings from academic literature, like the summary by Lüthje/Herstatt (2004) and the lead user project handbook by Churchill et al. (2009). Thus, her article serves as an appropriate reference to reflect the practical and academic state of the art.

Table 28 SOTA Lead User Method and the Proposed Method

	Eisenberg (2011)	Operational steps			Preference-Driven Lead User Method
Phase 1	Preparation to launch the lead user project	Team foundation	1.1	Team foundation	Preparation to launch the lead user project
		Project schedule	1.2	Project schedule	
		Master project plan	1.3	Master project plan	
Phase 2	Identification of key trends and customer needs	Trend analysis	2.1	Trend screening	Identification of trends and internal contributions
		Lead user classification	2.2	Definition of attribute set	
		Observation of needs	2.3	Identification of internal contributions	
Phase 3	Exploration of lead user needs and solutions	Lead user identification	3.1	Lead user classification	Exploration of the user community
		On-site visits and observation	3.2	Survey design and LU identification	
		Development of preliminary solution concepts	3.3	Survey distribution	
Phase 4	Improvement of solution concepts with lead users and experts	Administration and invitation	4.1	Data quality	Survey and preference measurement
		Concept refinement	4.2	Preference measurement state-of-the-art attributes	
		Business case	4.3	Analysis of internal contributions	
		Distribution to senior management	4.4	Collaborative filtering external contributions	
Phase 5	-	-	5.1	Pre-assessment of contributions	Lead user workshop
		-	5.2	Workshop preparation	
		-	5.3	Concept generation	
		-	5.4	(Optional) Re-test	
			5.5	Senior management go/stop	

4.3.   Detailed Description of the Phases

4.3.1. Phase 1 – Preparation for Project Launch

The 1st phase of the Preference-Driven Lead User Method describes the preparation of the project and starts with the *project grounding*. This is done in reference to the lead user project handbook (cf. Churchill et al. 2009) and is aiming to generate the master project plan.

Firstly, the process of *team building* (operational step 1.1) is fostered (see e.g. Stahl 2007 for a brief summary of the team building process). It is crucial for the progress of the project that the stages of forming, storming, and norming are passed quickly to perform teamwork (see Tuckman/Jensen 1977 for stages of small-group development and Bonebright 2010 for a review from an academic and a practical perspective). This becomes challenging in lead user projects since a diversified – interdisciplinary – team structure is recommended. This includes economic staff and internal technicians who may face NIH on a personal level, e.g. by rejecting the idea of user innovation resp. CAP in general. The interdisciplinary team is supposed to bring in diversified methodological knowledge but may face an extended storming process (see e.g. Gilley et al. 2010 for a summary of theories, a discussion on diversification, and its influence on the team building process).

Secondly, the *project schedule* (1.2) summarises the project focus and overall project objectives. In addition, resource requirements are incorporated in this phase. On the one hand, the project schedule describes the variables of a future product category, possible target markets and applications of interest, as well as expected business goals and an expected financial turnover. The necessary information is provided by the senior management. On the other hand, it describes the required human resources; time, and budget (cf. Churchill et al. 2009).

Thirdly, the project grounding also includes the orientation of the team to frame a common understanding of the innovation task, the applied methodology, and the approach used to proceed further. This describes the *master project plan* (1.3).

- *Precondition: Innovation task, project expectations*
- *Process: Team building, project definition, knowledge normalisation*
- *Postcondition: Master project plan*

4.3.2. Phase 2 – Identification of Trends and Internal Contributions

The 2nd phase of the Preference-Driven Lead User Method describes the *identification of key trends* and internal contributions as key aspects (see von Hippel 1986 and subchapter 2.3.4 for a discussion of trend evaluation being a major challenge and Franzen 1995 for a basic introduction to the topic). This describes the screening process for trends and is further fostered on expert interviews, patent research, and screening of literature among other possibilities.

Firstly, *trends are characterised* (2.1) as social, economic or technical streams (see Reichwald/Piller 2009, p. 183), but literature is ambivalent in defining the trend term. Typically, the term covers a constant development of one or more variables in relation to a given time period (see e.g. Müller 2008 in reference to Buck et al. 1998). This includes multiple perspectives that might be usable within the lead user method. Those are in reference to von Hippel (1986), Urban/von Hippel (1988) and Herstatt (2003): (1) market trends, (2) technology trends, (3) economic streams, (4) legal perspectives, and (5) social developments.

A *PESTEL analysis* (Political, Economic, Social, Technological, Environmental and Legal, see e.g. McGee et al. 2005 for the basic PEST analysis) can be a properly-structured process to identify trends. Sources of information can be categorised into analysis of secondary literature, expert interviews, and field reports.

The usage of *secondary literature* is, for example, performed and described by von Hippel et al. (1999), Herstatt et al. (2001), and Lüthje/Herstatt (2004). This covers a scan of magazines, scientific and practical journals, and trend reports (see for example Poetz et al. 2005, Churchill et al. 2009 and Sänn/Baier 2012 for practical insights). Visiting *exhibitions and fairs* may extend this.

Olson/Bakke (2001) report the use of *internal and external experts* for trend identification. This should be extended to experts from analogous markets (von Hippel et al. 1999) and by trend- and market researchers (Wagner/Piller 2011).

Heterogeneous expert wisdom and dissimilar evaluation structures (see e.g. Kornish/Ulrich 2011) demand a wide-ranged innovation focus to find real major trends (Lüthje et al. 2003). Poetz et al. (2005) showed an average of 33 expert interviews per lead user project that will lead up to 20 interesting trends to pursue further. Experts can be evaluated based on their impact both in academia and practice that may be derived from citations (see e.g. Hienerth et al. 2007). In contrast, literature shows discussion on the value of experts (e.g. Ozer 2009 in comparison with lead users) and boundaries that are attached with relying on experts solely (see e.g. Hoch 1988). In sum, data from experts may vary in relation with real consumer behaviour and product choices.

The lead user project handbook recommends a *trend investigation workshop* (Churchill et al. 2009, p. 76) to prepare an upcoming interview of experts. Field reports (von Hippel et al. 1999) describe another way to identify trends. This approach is resource demanding, but allows generating insights into the actual usage of a product. Gassmann/Gaso (2004) suggest alternative trend scouts.

The Preference-Driven Lead User Method does not recommend a specific methodology to identify trends and links to recommendations in Churchill et al. (2009). However, an *evaluation of trends may be skipped*, since respondents will be able (1) to provide feedback on trends in the later survey and (2) the underlying trend tends to be pre-defined by the master project plan w.r.t. the innovation task that is given by the senior management (in practice). In addition, recent developments in the field of crowd-based innovation point to beneficial input from idea markets for trend identification (see e.g. Natalicchio et al. 2014 for a literature review), too. Thus, the Preference-Driven Lead User Method distinguishes between trend evaluation and *trend screening*. It focusses on the latter one and handles trends as initial values that can be extended by respondents in the later survey. This shifts trend evaluation from the project team to the respondents, since each additional trend can be expressed as a new category within the later questionnaire. Moreover, this allows screening for trends that were not visible for the project team in the first place. Literature has further shown that the evaluation of trends is in fact based on the feasibility of a trend in relation to the innovation focus and on the presence of the trend in secondary literature or by experts. Moreover, Herstatt/von Hippel (1992), Poetz et al. (2005), and Hienerth et al. (2007) have shown that multiple trends (three and more) were pursued for a later development in lead user projects anyway.

Secondly, state of the art market available solutions are gathered and decomposed to their *attributes and attribute levels* (see chapter 3.2) that define an attribute set of product components (2.2) and detail the later survey design. The composition of the attribute set is typically done with professional assistance (in-house or external) since the combination of different attribute levels with each other shall be makeable, meaningful, and realistic (see Baier/Brusch 2009 for an overall description of requirements for building attribute sets).

Thirdly, the development and *identification of internal contributions* by engineers and/or developers is the next task (2.3). The main purpose is to overcome – or at least to bypass – the NIH and to generate raw product ideas as a starting point for external ideation (see Katz/Allen 1982 for the basic definition and Poetz et al. 2005 for possible aspects on how to lower the NIH).

The interdisciplinary team is allowed to *contribute with own ideas and concepts* w.r.t. the project focus and the already discovered trends. This reflects the internal knowledge within the companies' departments and can be further stimulated by internal (lead user) idea competitions, internal signaling or by order (see e.g. Schweisfurth/Raasch 2015 for lead users inside the firm).

The raw ideas and concepts derived will be *included in the later survey*. Fundamentally, the gathered input from this additional ideation can broaden the team's understanding of the state of the art and stimulate internal discussion. This can further reflect a decision-making process of a buying centre in business-to-business applications and sharpen the project team's sensibility for the target market in general (see for example Webster/Wind 1972 for the institutional buying behaviour as reference and von Hippel 1986).

However, the team must also perform a *trade-off between the amount of information in the survey and the survey complexity* (see in reference to chapter 3). This influences the method for the later preference measurement and the design of the survey in the prototype of the Preference-Driven Lead User Method.

- Precondition: Master Project Plan
- Process: Trend screening; analysis of state of the art components, definition of attribute set, internal ideation
- Postcondition: Method of preference measurement, survey design

4.3.3. Phase 3 – Exploration of the User Community

The 3rd phase of the Preference-Driven Lead User Method covers the specification of lead user characteristics (*lead user classification*) to prepare the questionnaire that is employed to explore the user community.

Firstly, the *lead user classification* (3.1) can be achieved in multiple ways (see subchapter 2.2.3). The dimensions of "ahead of the trend" and "high-expected benefit" can be used to measure the latent construct of lead userness. Relevant questions to build these dimensions were adopted from Batinic et al. (2006), Franke et al. (2006), Faullant et al. (2012), and Schweisfurth/Raasch (2012). Literature shows sufficient results for lead user classification solely based on *trend and benefit indicators* (see e.g. Faullant et al. 2012 and Schweisfurth/Raasch 2012), too. The expected benefit expresses a form of dissatisfaction, which is proven to possess significant influence to increase idea quality (see Schuhmacher/Kuester 2012 for empirical evidence in new service development). Table 29 illustrates suggested items of both dimensions that are employed to generate the lead userness construct in the prototype of the Preference-Driven Lead User Method. A 6-point rating scale is chosen in the prototype to rule out an in-the-middle option. Literature has not agreed on a specific scale, but tends to prefer the Likert scale.

Table 29 Exemplary Items for Lead User Classification

Construct	Items for High Expected Benefit (LUBE) and Ahead of Trend (LUAT)[*]
LUBE1	I frequently face problems concerning mobile application of mass spectrometers, which are not solvable with current products on the market.
LUBE2	I am dissatisfied with some components of available mobile mass spectrometers.
LUBE3	Manufacturers of mobile mass spectrometers were not able to help me with application problems in the past.
LUBE4	In my opinion, there are unsolved problems with current mobile mass spectrometers in detail.
LUBE5	Mobile solutions that are available on the market do <u>not</u> fit my personal expectations on mass spectrometry.
LUBE6	The low user orientation of mobile mass spectrometers manufacturers disturbs me greatly.
LUAT1	I actively follow current developments on the market of mobile mass spectrometers.
LUAT2	I have already gained benefit from the early use of new developments in mobile mass spectrometry and prototypes.
LUAT3	I participate in the usage of mass spectrometers under extreme conditions.
Items were measured on a 6-point rating scale: 1... totally disagree; 6 ... totally agree	

Additional items to explain aspects like experience, technical knowledge, and innovativeness in general are implemented in an optional part of the survey to sharpen lead user classification if necessary. This remains optional in favour of an increased usability for the respondent.

Chapter 2 shed light on the discussion of using *own idea development*, leading edge status or lead userness to characterise lead users. Those items were adopted from Poetz/Schreier (2012), Schreier/Prügl (2008), Franke et al. (2006), and Herstatt/von Hippel (1992). The most important aspect in the prototype of the Preference-Driven Lead User Method is a direct questioning to identify an (2) own idea development that can be used as a sufficient indicator for lead user classification. This is derived from previous experiences in multiple pre-test studies (see Sänn et al. 2013, Sänn/Ni 2013, and Sänn/Baier 2012).

Thus, this 4th part of the questionnaire asks for *previous participation* in product development, the addressed problem, and the related trend as well as the achieved development level. The generation of external contributions is based on this idea development. The ability to *stimulate ideas* results from the basic assumption that idea markets are able to stimulate ideas based on the presence of other ideas (see Soukhoroukova et al. 2012 for using this ability as an evaluation criterion).

The applied generation of external contributions expresses a problem-focused view, but it can be converted into a *solution-focused view*, too. The latter one will show contributions in form of a solution to a specific need. Solution-focused input should be stimulated primarily, since this input includes the problem, illustrates an already approved approach, and provides greater value for further NPD (see Mahr/Lievens 2012 for a discussion of problem-focused and solution-focused contributions in the case of new service development). Respondents further decide whether they like to share the idea with the community or not.

Secondly, the decision to follow a specific *identification process* (see subchapter 2.2.3) fosters the later survey design (3.2). The choice of the identification process depends on the available resources and the project team's willingness to reveal the innovation project to the public or to selected respondents, exclusively. The prototype uses screening as the identification process (see subchapter 4.2.3 for a short discussion on handling alternative identification processes).

The *survey design describes aspects of preference measurement*, item positioning, and graphical design. Survey design, preference measurement and lead user identification depend on each other. The prototype template for preference measurement employs traditional self-explicated measurement for complex products and adaptive conjoint analysis for simple products. Findings of

Eckert/Schaaf (2009), Sattler/Hensel-Börner (2007), and Chrzan/Golovashkina (2006) support this approach in terms of usability, reliability and validity.

The *design needs to further incorporate motivational aspects* (see table 7) to stimulate participation. For example, an additional motivational factor that needs to be highlighted is the later participation in the workshop that can enhance one's own reputation and promises access to new sources of information for the respondents. The overall response rate and response quality can be increased by using incentives (see Selka 2013 for a basic summary and subchapter 2.2.5).

The structure of the questionnaire in the Preference-Driven Lead User Method includes an introduction to explain and to promote motivation and participation. The main framework of the survey includes: (1) lead user classification, (2) preference measurement, (3) evaluation of internal and external contributions, (4) idea stimulation, (5) validity measurement, and (6) additional data that covers the respondent's experience in the application field, their profession, and their job position within the company's buying centre for a business-to-business application. The latter one frames the closing part of the survey and concentrates on demographic data and other time-consuming questions. In addition, (7) an option for feedback to the survey is supplied next to the possibility to share and recommend this survey by e-mail and on social networks. Overall, the graphical design is customisable and dependent on the favoured evaluation techniques of internal and external contributions, the target audience and their primary input device to answer this web-based survey.

Thirdly, the *survey distribution* (3.3) depends on the available resources in terms of time, human resources and budget. The search for lead users and non-lead users should cover the target market and analogous markets to overcome the local search bias (see subchapter 2.2.4). The prototype framework of the Preference-Driven Lead User Method suggests a *hybrid approach* and uses screening techniques coherent with pyramiding.

Screening is employed to promote the survey via online discussion boards, websites, social networks, and by e-mailing to previously generated contacts and already existing business partners in the target market (see Herstatt/von Hippel 1992 for a similar approach in the Hilti project). This leads to the benefit of a fast market-wide distribution of the survey and a fast response to overcome the cold start problems of collaborative filtering and idea stimulation. The integration of multipliers like lobby associations is recommended to achieve a maximum market coverage. Manfreda et al. (2008) and Bosnijak (2013) found that the average response time for web-based surveys by respondents with mobile devices is about six to eight hours.

The parallel or later pyramiding part starts with previously identified experts as the bottom of the pyramid. The pyramiding approach uses the questionnaire to verify the lead user status. It further links personal recommendations to address respondents with an increased level of lead userness and an advanced experience in the target market and in analogous markets (see Poetz/Prügl 2010 for benefits of pyramiding to achieve a fast shift to analogous search fields). This is integrated in the optional part of the questionnaire. Further conducted *distribution at special events*, e.g. at industry fairs, exhibitions and conferences, may also be possible. However, a maximum coverage of the target market (population) is a basic driver next to a focus on *quantity in idea stimulation*.

The project team acts as a *moderator* during the survey distribution. Incoming external contributions need to be scanned for their meaningfulness, correct grammar, wording, and for duplicates. This includes translation to the voice of the ordinary customer (see Griffin/Hauser 1993). The team needs to approve the contributions and to mark them as visible for other respondents. The respondents can then receive feedback on their contributions from the project team and may also be interested in further feedback by other respondents. This is a significant motivational anchor (see table 7) to stimulate external contributions and the distribution of the questionnaire.

- Precondition: Master project plan, attribute set, internal contributions
- Process: Lead user classification and identification, survey design, distribution
- Postcondition: Survey data, list of external contributions

4.3.4. Phase 4 – Survey and Preference Measurement

The 4th phase of the Preference-Driven Lead User Method describes the analysis of preference data after survey accomplishment. The evaluation of contributions is parted into (1) the analysis of state of the art attributes and (2) the evaluation of internal and external contributions.

Firstly, *data quality* needs to be proven (4.1). Thus, lead users need to be identified by the lead user classification. In the case of lead user classification by an own idea development, this becomes dichotomous. Classification based on the lead userness construct requires an in-depth analysis. Cronbach's alpha, explained variance, item-to-total correlation, and validity criteria were seen as relevant *quality requirements* to depict adequate questions for characterisation of lead users (e.g. Franke et al. 2006). Cronbach's alpha measures the internal consistency (congruency) of the underlying questions that were used to compose a dimension of the lead userness construct and thus can serve as a quality filter here.

One would assume that the highest rating in classification questions would lead to an identified lead user. This is not true for every case though, e.g. if a respondent simply tends to rate high values or just clicks through the survey. Thus, an *own idea development serves as a validator*. In sum, the project team needs to *separate between lead users and non-lead users*, analyse preferences for the given attributes and attribute levels per user group and jointly (overall).

The resulting part-worths indicate preferences. The *validity* of the preference measurement is given by e.g. a correlation between the results of this analysis and the results of the holdout-part of the questionnaire. In general, the calculated part-worths for the attribute levels are added to reflect the stimuli in the holdout set. The direct evaluation, e.g. by ranking the holdout cards in accordance to the respondent's preference, is compared with the generated evaluation of the stimuli to proof (external) validity. When ranking is applied, one would use the Spearman correlation factor. *Reliability* can be proven by using an (optional) re-test under equal conditions after the workshop.

Secondly, *preference measurement* for state of the art attributes and attribute levels (4.2) is fostered on standard methods (as introduced in chapter 3). It is recommended to distinguish between the lead user group and the non-lead user group to get first insights into their *preference structures* (in reference to Urban/von Hippel 1988). Statistical tests may serve to detect significant differences in the evaluation of the attribute levels per user group. The result points to similar evaluation structures or not. *It is to assume that the project team does not want to have a similar structure in this case* and it can be argued by the fact that actual products do not serve as a satisfactory solution for lead users. Otherwise, this would point to (1) a wrong lead user classification – a misleading characterisation of lead users –, (2) a wrong lead user identification – no lead users were found or only lead users were found –, (3) or a missing portion of ordinary customers – that may be an unwanted side-effect of pyramiding.

Thirdly, binary measurement is used for the *analysis of internal contributions* (4.3). Mean values, respectively the overall count, point to contributions that need to be taken into consideration and reflect the (overall) acceptance by the respondents. As mentioned previously, detailed preference values might be biased and may lead to the exclusion of promising contributions in this early stage of product development. Internal contributions can be analysed in a direct way, since every respondent had to rate them. The project team needs to distinguish between lead users and non-lead users, to understand the preference structures, respectively the evaluation structures of both (in reference to Urban/von Hippel 1988). It is assumed that the project team is looking for similar evaluation structures to identify promising internal contributions that are preferred by both user groups.

Fourthly, *collaborative filtering* allows rendering evaluations of external contributions (4.4). The fundamental approach of the Preference-Driven Lead User Method to incorporate additional data on the fly leads to an incomplete evaluation matrix of external contributions. This would lead to a monotone pattern for missing values in an idealised system (see table 30 in relation to Horton/Lipsitz 2001, p. 245 figure 1). Respondent (user) u_1 suggested external contribution e_1 to the survey that was approved by the project team and included in the questionnaire. Respondent u_2 is then able to evaluate e_1, contributed e_2 and so on. Unfortunately, respondent u_1 is not able to evaluate e_2 by design, but the following respondents will be able to perform this evaluation task.

This leads to a basic data set for external contributions as given in table 30.

Table 30 Incomplete Evaluation Matrix of External Contributions

Item/User	e_1	e_2	e_3	e_4	e_5	...	e_n
u_1	x	-	-	-	-	-	-
u_2	1	x	-	-	-	-	-
u_3	0	1	x	-	-	-	-
u_4	1	0	0	x	-	-	-
u_5	1	1	0	1	x	-	-
u_6	0	1	1	0	1	-	-
u_7	0	0	1	1	0	...	-
...	...	...	...	...	...	...	...
u_m	1	1	1	0	1	...	x

As a result, the project team receives a list of external contributions with the estimated importance of each specific contribution for each respondent. A cluster analysis of external contributions can point to heterogeneity or homogeneity.

- Precondition: Survey accomplishment, binary values for contributions
- Process: Data quality, preferences structures, acceptance values, cluster
- Postcondition: Favoured internal and external contributions

4.3.5. Phase 5 – Lead User Workshop

The 5th phase describes concept development within a workshop setting and an optional re-test. It is supported by the pre-assessment of gathered contributions.

Firstly, a *pre-assessment of internal and external contributions* in relation to available resources, feasibility, and production is done (5.1). This prepares the upcoming workshop and addresses the effectiveness of the lead user method in general. The associated NIH (see Antons/Piller 2014 for summary of the state of the art) may occur, e.g. in forms of type 1 or type 5, and leads to a biased assessment of external contributions in these terms. However, this NIH may be deflated by the presence of acceptance values and this leads to a principle "When in doubt of a contribution, then include the contribution for further processing".

The *pre-assessment* allows the application of multiple evaluation techniques like the function point method (see e.g. Sharma et al. 2014 for an overview). Further dimensions of an appropriate evaluation can be derived from QFD (see e.g. Griffin/Hauser 1993). This specific step serves (1) to *evaluate the feasibility of current concepts*, (2) to *prepare an implementation of contributions* and (3) to *avoid resource-based miscalculation*. The project team may be encouraged to incorporate further internal engineers and developers to assess the effort for concept development. This provides feedback of market segments and allows calculating the costs of large-scale production (see Hienerth/Lettl 2011).

On the one hand, internal and external contributions with *high acceptance values and low transformation costs* will be chosen to cover a broad market. This leads to a lead user workshop with a *focus on concept development*. On the other hand, the project team can choose internal and external contributions with high acceptance value and high transformation costs. This may lead to a lead user workshop that focuses on the *development of new processes* to make the promising contributions feasible to be produced by the present infrastructure.

Secondly, the *preparation of the workshop* includes the selection and invitation of participants who provided favoured contributions (step 5.2; see e.g. Gottesdiener 2003 for recommendations related to the workshop and the team building process within the workshop). Insights from practice have shown that the German law may complicate a standard workshop process due to the *ArbnErfG* (employee invention law) which is especially applicable for business-to-business contexts. The general management has to take care of that challenge by negotiating about IP-related agreements or by holding the workshop in a foreign country. Reports from 3M described that workshop participants were willing to sign over "... any property rights that might result from the workshop" (von Hippel et al. 1999, p. 53). Recent literature highlights the relevance of a fair commercialisation and thus a free sign over of contributions remains doubtful (see Franke et al. 2013).

Thirdly, the further focus lies on *concept generation*. The workshop details the product concepts (5.3), reveals insights for necessary development tasks, and allows the company to get familiar with its lead users. Previous recommendations suggest to frame the workshop based on structured analysis to decompose favoured contributions (see DeMarco 1979). This is consistent with the traditional lead user workshop (see Herstatt/von Hippel 1992 for a detailed description).

Fourthly, the (optional) *re-test* of the generated evaluation of external contributions and of the final concept depends on the experience of the project team and their estimation of the goodness to fit relevant market requirements by the favoured contributions (5.4). The re-test with a final product concept *validates the results of collaborative filtering* and provides *additional data on reliability of preference measurement*. This can be beneficial to allow a more sophisticated prediction of future market shares. The 1st presented stimulus is a market-available product (e.g. from a major competitor or a niche development, see for example Sänn et al. 2013). The 2nd holdout describes the featured concept from the Preference-Driven Lead User Method. The 3rd option is either a minimal or maximal set of possible attributes or another reference product. The measurement of reliability for the lead user classification can be done here, too. Empirical studies have shown that market distance (analogous markets) and external technical knowledge positively influence the novelty of the resulting concept but decrease feasibility (e.g. Hienerth et al. 2007). Thus, the re-test may also be accomplished in *analogous markets*.

Fifthly, the outcome of this workshop will be distributed to the *senior management* level that will decide whether to continue the development or to stop the project (5.5). This is a standard progress adopted from the traditional lead user method and is a crucial point that decides whether the project was internally successful or not. The Preference-Driven Lead User Method supports the senior management and provides the final concept, an estimation of necessary resources, and preference values to emphasise feedback from the market. This should avoid an expensive development of a new product that addresses a niche market and faces a long-term diffusion process. Overall, this can support the decision-making process and may lower the uncertainty to apply the method. Additional data to support the project would be generated in the (optional) re-test. The adjourning stage of the team process is further reached.

- Precondition: Preference and acceptance data, filtered contributions
- Process: Assessment of effort and feasibility
- Postcondition: Final concept, resource requirements for further development, evaluation structures, validated input, go/stop decision

4.3.6. Implementation

Version 1.92+ (Build 120330) of LimeSurvey provided the software environment to realise the prototype concept of the Preference-Driven Lead User Method.

LimeSurvey is an *open-source software tool* to develop standardised online surveys and to customise them in multiple points under the conditions of the GNU General Public License (GPL) in version 2 and provides the possibility to *implement additional features*, which are required to make the Preference-Driven Lead User Method work. In contrast to other software distributions for designing online survey, LimeSurvey is not developed as a software-as-a-service application (SaaS) and is self-hosted on a web space using PHP and MySQL. *JavaScript* is used on the client's browser. The resulting survey data is provided in multiple formats to be further processed in e.g. Microsoft Excel, SPSS, and R. Schmitz (2012) provides an overview with detailed information about LimeSurvey.

Firstly, the *identification of additional contributions* is placed in the 4th part of the survey that is also known as idea stimulation (subchapter 4.3.3). Either the respondents provide information about the development project they have been involved in or they provide insights into a basic problem that was encountered. The gathering of contributions is detailed by questions about the application field and the problem area. The development level of the contribution determines if the ideas were shifted to a further status, like design concept or prototype. If respondents provide information about an own development work and want to share this, then the contribution is saved in an array with the user id. In the prototype, this stores only further developed contributions, but is customisable.

If one user is willing to share the contribution, then a *notification is send to the project team* using a RSS 2.0 feed (Really Simple Syndication, see the RSS advisory board for its specification). RSS feeds are XML formatted plain texts and allow sending notifications on events using an information push approach. Members of the project team add the address of the RSS feed (e.g. http://prelead.de/methods/rss.php) to their standard e-mail clients or web browsers and receive a notification on new user contributions. Typically, the notification is done just in time.

Notification messages include a title, an idea description, the publication date and the userID. A new node that describes the new contribution is added to the XML tree. This is *triggered* when a respondent passes the 4th part of the questionnaire using the "next" button. The project team receives a message and can proceed with their moderation task. The related pseudo-code is given below.

READ UserID

READ Contribution

IF Content of External Contribution is not empty AND Level of External Contribution is greater than 2 AND Willingness to Share External Contribution is TRUE

THEN Append (UserID, Content of External Contribution, Category of Contribution) to Array

Append UserID and Content of External Contribution to XML tree as node Item

UPDATE RSS stream

Secondly, the *moderation task* includes translation to the ordinary VoC, if necessary, and the decision to publish this contribution to future participants. Members of the project team log in to LimeSurvey to *check the contribution for pre-defined criteria*. After the contribution is verified, it can be taken into consideration by future analysis within the 3rd part of the questionnaire. Pre-defined criteria can be correct wording, a comprehensible description and uniqueness in relation to other contributions within this survey. The function to verify a contribution is given in LimeSurvey by using a workaround with an additional checkbox. This is checked by the project team to publish a contribution.

Thirdly, the *integration of user contributions* is applied in the 3rd part of the questionnaire, which is 'evaluation of internal and external contributions' (see subchapter 4.3.3). This is done by reading every input from the problem/solution question of the survey.

The read operation uses an *SQL query* to select every entry from the survey database that matches the criteria of being shared, further developed, and released by the project team. The idea text and the category of the problem field (that can reflect a trend) are extracted for each entry. The results of the query are parsed and stored to an associated array that is returned to JavaScript.

JavaScript on the client's side *identifies each contribution* using the internal ID of the associated question for the contribution's text and places it after the internal contributions of a specific problem field (also referred to as trends and categories). The placement can also be done randomly to avoid order effects.

The *presentation of internal and external contributions* in the user front end starts with the positioning of headings for broader system components, problem fields, and trends. A-priori known problem fields and trends are associated with internal contributions and LimeSurvey starts by listing all internal contributions. JavaScript adds corresponding headings to each 1st contribution of a specific problem field. Next, external *contributions are inserted in association to the correct category* headers. Contributions are added with checkboxes in the present implementation of binary evaluation. The related pseudo-code is given below. It is apparent, that in theory the number of attributes, problem areas, and trends can *easily exceed a respondent's cognitive capacity* and thus challenges derived from preference measurement may occur (see chapter 3.5). This emphasises the importance of the moderation task and the necessity of using collaborative filtering (see chapter 3.4).

CONNECT to SQL Database

SELECT DISTINCT (Category of Contributions)

FOR Categories n = 1 to m

    Define string variable n-th_header for Category of Contribution n

    FOR Internal Contributions a = 1 to b

        IF Category of Internal Contribution a equals n-th header THEN

        Prepend n-th header before Internal Contribution a AND Set a=b

    FOR Internal Contributions a = 1 to b

        IF Category of Internal Contribution a equals n-th header THEN

        Append Internal Contribution a after n-th header

SELECT (Content of External Contributions AND Category of Contribution)

WHERE Level of External Contribution is greater than 2 AND

Willingness to Share External Contribution is TRUE

FOR Categories n = 1 to m

    FOR External Contributions a = 1 to b

        IF Category of External Contribution a equals n-th header THEN

        Append External Contribution a AND Add Checkbox for a after n-th header

Fourthly, the user front end for *evaluation and analysis of internal contributions* is provided by LimeSurvey. Ratings for *external contributions* are handled using an auxiliary variable, since the underlying SQL structure is fixed and LimeSurvey does not allow adding more items in an active survey.

Each checkbox for external contributions is set to be false per default. Elsewise, respondents have to decide whether they like a contribution or not by selecting the related checkbox for every item. This evaluation is assumed to complicate the evaluation task and may be *time-consuming*. The related pseudo-code for the prototype of the Preference-Driven Lead User Method is given below.

FOR External Contribution n = 1 to m

    IF Checkbox is "FALSE" THEN {

      WRITE Content of External Contribution n and "FALSE" to Array}

    ELSE {

      WRITE Content of External Contribution and "TRUE" to Array}

Hahsler (2011) noticed for analysing ratings that a default value "false" is an indicator *either that the respondent did not notice the contribution or rejected it* (see subchapter 3.4.1). Non-available contributions are presented as NULL in the final evaluation matrix. The resulting respondent-contribution matrix is exported by default functions of LimeSurvey and will be imported in R for further processing.

4.4. Application in Mountain Biking

4.4.1. Preface of the Story

This subchapter presents an application example of the Preference-Driven Lead User Method in the field of mountain biking that was initiated in February 2011. This application is a follow-up work of Sänn/Baier (2012, 2014) in an online environment and was adapted to fit the Preference-Driven Lead User Method in detail. The application setting is framed at the fictional company MountainCross.

MountainCross is a mountain bike manufacturer from the Black Forest region in Germany and frames the application field. It can be described as a small-sized enterprise (according to the definition of the European Commission 2005). It is well-known for its handmade frames and for using reliable mountain biking components. However, MountainCross faces financial challenges due to serving a niche market and a previously failed opportunity to register their bicycle design. Now, a major competitor offers a similar design using automated mass production for a lower price. MountainCross' target group is known to have a sense for build quality, but is also aware of the price and therefore the company's future is in doubt. Summarised, MountainCross faces multiple challenges that were found using an analysis of its strengths and weaknesses (see table 31 in reference to Moser/Piller 2006 for a comparison).

Table 31 Analysis of Strengths and Weaknesses

	Internal Strengths	Internal Weaknesses
External opportunities	Increasing awareness for high product quality and reliability by customers	IP protection, market growth with limited resources
External threats	Handcraft assembling processes needs to be adapted to automation	Increasing price sensitivity and global competition

Fortunately, the CEO is aware of the situation and expects to regain a leading market position while facing a broader market by the development of new products. MountainCross' employees can be characterised as bicycle enthusiasts with a passion to perform exacting mountain biking activities, while having strong ties in the community. Thus, it is expected that there are valuable ideas inside the company. Nevertheless, the heterogeneity of the target group leads to general uncertainties to rely on internal contributions, solely.

4.4.2. Phase 1 – Preparation for Project Launch

The project starts with *project grounding*. The *project team* (1.1) consists of an innovation manager, a member from accounting, and two technicians from preproduction and final assembly. A further sales agent and an employee from procurement, who described himself as a passionate mountain bike cyclist, were included in the team. The overall *task to innovate* was given by the CEO and the general market of cycling was defined as the target market (1.2). The background of the innovation task aside with its resource-based constraints is explained in a kick-off meeting. A common understanding of the applied method is established by a presentation and led to an on-going discussion about external input.

This project is summarised in the *master project plan* (see table 32) covering product category, target markets, applications of interest, business goals and constraints (1.3). The plan can be edited in later phases of the development process since it is expected that the project team will learn about the market with trends and opportunities as well as about analogous markets.

Table 32 Master Project Plan for the MountainCross Case
(Reference: Illustration in Reference to Churchill et al. 2009)

Scope	Definition
Product category	We are seeking to develop a new concept for a mountain bike to cover an extended target group of individuals with high quality standards to foster our sustainable existence.
Target markets	End users: 'hipster' in the middle price segment with high quality standards; current distributors: specialist retailers; new distributors: mass retail stores
Applications of interest	Business applications: e.g. Bike messengers, pizza services, postal services, police; sports: semi-professional cyclists
Desired project outcomes	Identification of market/need areas that represent strong outcomes; development of one concept for a mountain bike that can be brought to market within the next 1 or 2 years
Key business goals & constraints	Revenue EUR 76 million for the 5th year after market introduction; concept should utilise current components; product should enable continued reliance on current partners

4.4.3. Phase 2 – Identification of Trends and Internal Contributions

The *identification of trends* (2.1) starts with research by gaining information from several magazines like Fahrrad News and mtb. In addition, commercial catalogues that were published by competitors like Giant, Stevens, Specialized, and Steppenwolf are used to show market-available configurations.

Further, basic literature on patents in the biking field is scanned (see Herzog 1991 for a detailed listing on historic bicycle patents) to become aware of already existing concepts. Surprisingly, historical patents show congruence with current innovation streams, like the ebike concept. In addition, rebiking is found to be a rising trend.

These streams were validated by *scanning online discussion boards and interviewing (local) experts*. User comments on MountainCross' products were further reported to be beneficial to this project. In addition, multiple fan-websites are analysed (see e.g. Genesbmx 2014). This reveals state of the art products and leads to mind-opening insights to future fan-based cycle designs in general.

The overall findings are aggregated in three runs to define a relevant *attributes set* (2.2). The 1st run covered 51 interesting trends and derived attributes, among them attributes like: (1) brakes, (2) ergonomics, (3) colour, (4) suspension, (5) weight, (6) wheels and tires, (7) transmission, (8) maintenance, and (9) further accessories. The identified attributes led to various ideas to reduce weight, increase stiffness, improve reliability, and to increase comfort. The 2nd run covered 19 attributes and the 3rd run resulted in five attributes with three resp. two attribute levels. All of these were market-available attributes and levels to reflect possible state of the art stimuli. Table 33 provides an overview of the selected attributes and their levels.

Table 33 Selected Attributes and Levels for Preference Measurement

(Reference: Illustration in Reference to Sänn/Baier 2012)

Presented Attributes	Standard Level	Incremental Level	Breakthrough Level
Suspension	No suspension	Hard tail suspension	Full suspension
Transmission	Standard drive train	Carbon chain guide	Gear belt
Wheels & tires	Standard wheels & tires	Run flat tires	Fiberglass wheels
E-bike concept	No, w/o power assistance	Yes, with individual assistance	
Safety	No, w/o pedal lock	Yes, with pedal lock	

The employed attributes for preference measurement covered parts of the *suspension, transmission*, wheels and tires, the e-bike concept and a *safety attribute* – called the "pedal lock" that was derived as a lead user contribution from previous literature (see Lüthje et al. 2005). Incremental attribute levels were market available and could be applied with minor changes to a standard concept of MountainCross. Breakthrough attribute levels were also available, but demanded major changes to a basic concept. These were used in other application fields like in indoor cycling with great success. Overall, the resulting attributes and their levels are incorporated to the survey by using alternative dictions in the voice of the ordinary customer with an adjusted information load.

The *identification of internal contributions* (2.3) is derived from in-house input (here it is substituted with findings in a previous survey, see Sänn/Baier 2012). This revealed multiple contributions covering various trends and innovation aspects, among them adaptive seating, design adaptions, a citykit, a novel frame concept, frictionless light, a novel maintenance concept, and a *smart braking system*.

4.4.4. Phase 3 – Exploration of the User Community

The *lead user classification* (3.1) is done using an own idea development and the lead userness construct. The following statements were derived from literature (Sänn/Baier 2012 in reference to Lüthje et al. 2005) to verify the classification of lead users and non-lead users by previous experience and is given in table 34.

Table 34 Aspects of Lead Userness for MountainCross

(Reference: Illustration in Reference to Lüthje et al. 2005 and Sänn/Baier 2012)

Dimensions of Lead Userness	Items Used to Generate Dimensions
Ahead of trend	I perform more than one discipline. I attend bicycle tournaments. I write experience reports.
High expected benefit	I train a certain skill. I had an idea of improvement.
Technical knowledge	I have technical knowledge about my equipment. I have knowledge about tools to repair my bicycle. I repair my bike on my own.

The *survey design* and the choice for a method of *lead user identification* (3.2) go hand in hand. The lead user identification shall be done using a screening approach by employing a web-based questionnaire, but also enable recommendations at the same time. Thus, the *hybrid approach* (see subchapter 4.3.3) is used.

Adaptive conjoint analysis (see subchapter 3.3.1) is used for the enclosed preference measurement of state of the art attributes and levels. Previous literature (see e.g. Helm et al. 2008) provided experiences in this field. The employed ACA is based on Sawtooth Software packages in this case, but various conjoint software packages exist on the market like LEET Conjoint for CBC.

The resulting questionnaire consists of seven parts. The 1st part of the questionnaire investigates the *customer's user experience* along with their *technical skills*. The 2nd part is ACA with state of the art attributes (see table 33). The 3rd part asks about internal and external contributions that were measured using binary rating. The 4th part asks about the self-made innovation, the innovation depth, and the basic problems the user addresses. It further examines the sources of information to develop the solution. The 5th part of validity measurement is inherited in ACA for internal validity and used holdout sets to prove external validity. The questionnaire finishes with demographic items (6th part) and an invitation to give feedback and make further recommendations (7th part).

The *survey distribution* (3.3) covers about 50 online discussion boards in Germany, Switzerland and Austria. Overall, about 500.000 registered users are theoretically reached using the screening approach to search for lead users.

The online survey was distributed in May and June 2011 with undergraduate assistance. Table 35 summarises the addressed online discussion boards.

Table 35 Online Discussion Boards and Survey Distribution

Online Discussion Board	Hosted in Country	Number of Participants	Sample Share in %
Bikeboard.at	Austria	8	7.7%
Gipfeltreffen.at	Austria	8	7.7%
Mtb-News.de (IBC)	Germany	19	18.3%
Mtb-Forum.eu	Germany	7	6.7%
Querfeldein-Kurbeln.de	Germany	4	3.8%
Fk-Riders.net	Germany	3	2.9%
Roberge.de	Germany	3	2.9%
Live-Radsport.ch	Switzerland	3	2.9%
Other discussion boards	DACH Region	49	47.1%
Overall		104	100%

4.4.5. Phase 4 – Survey and Preference Measurement

The online study generated 121 completed surveys out of 146 questionnaires (83%). The resulting sample for preference measurement of n=104 questionnaires met the *quality requirement of a minimum Spearman correlation* (r=0.5) and was used for further analysis (4.1). This was the result of the measurement of external validity and led to an average Spearman r=0.81 that indicates sufficient results.

The construct of lead userness was used to perform *lead user classification* (see table 29). The employed dimensions were "being ahead of a market trend" (Cronbach's Alpha=0.55) and "expecting a high benefit from a solution" (Cronbach's Alpha=0.71). Besides, this lead user classification linked positively to the dimension of "technical knowledge" (Cronbach's Alpha=0.70) (see table 34). Not all dimensions reached the sufficient value (≥0.7). However, the resulting lead user classification using the construct of lead userness covered 100% of an own – further shifted – idea development. In contrast, an own development covered only 70.4% of the lead users that would be revealed by lead userness. Thus, using an own idea development is regarded as acceptable.

Demographic and descriptive data shows that the average age of all respondents was 32 years, with 80% male and 20% female participants. About 76.9% of the respondents owned more than one bicycle. 81.7% of the respondents performed more than just one discipline and further 94.2% of the respondents said that they have technical knowledge about the equipment. Table 36 summarises this. Overall, 35 external contributions for the mountain biking field were collected which leads to a ratio of user innovation of 33.7% and is sufficient.

Table 36 Lead User-related Questions in the Survey

Descriptive Data and answers for Lead User Classification	Overall Respondents (n=104)
I own a mountain bike.	94.2%
I own another bicycle (BMX, racing etc.).	76.9%
I perform more than one discipline.	81.7%
I attend in bicycle tournaments.	35.6%
I train a certain skill.	57.7%
I have technical knowledge about my equipment.	94.2%
I have knowledge about tools to repair my bicycle.	91.3%
I repair my bike on my own.	85.6%
I had/have an idea of improvement.	33.7%
I write experience reports.	46.2%
Average usage time of the bicycle per week.	9.3 hours
Average use of a mountain bike in general.	21.9 years

The *analysis of preferences* (4.2) covered state of the art attributes. The resulting preferences are given below as part-worths for the whole sample (table 37). Overall, respondents prefer a full suspension that covers the front and the rear axle. Further, the overall results are characterised by standard equipment like standard transmission, wheels and tires. The pre-defined standard concept reflected a standard mountain bike from a middle class price segment in this survey. Table 37 illustrates the resulting concept and provides the overall ratings as part-worths. This is without any applied lead user classification.

Table 37 Results of Adaptive Conjoint Analysis in Mountain Biking

| Attribute Level | Part-Worth of Attribute Levels | Resulting Product Concept |
	Overall (n=104) Mean (Std. Dev.)	
No suspension	-1.810 (1.303)	
Hard tail suspension	0.739 (1.141)	Full suspension
Full suspension	1.071 (1.583)	
Standard transmission	0.387 (1.016)	
Carbon chain guide	0.105 (1.455)	Standard transmission
Gear belt	-0.492 (1.273)	
Standard wheels & tires	0.430 (1.153)	
Run flat tires	-0.303 (0.996)	Standard wheels & tires
Fiberglass wheels	-0.127 (1.354)	
No power support	1.373 (1.188)	No introduction of the
Power support	-1.373 (1.188)	ebike concept
No pedal lock	0.369 (0.974)	No pedal lock
Pedal lock installed	-0.369 (0.974)	

A further *separation of the results for a group of lead users and a group of non-lead users* is performed using an own idea development as a sufficient indicator. In sum, 35 respondents contributed with an idea, but only 24 were further developed. The development level was used to differentiate between lose thoughts and innovative contributions. Thus, 24 respondents were identified as lead users.

Table 38 Preferences in Relation to the User Group

| Attribute Level | Part-Worth of Attribute Levels by User Groups | | |
	Overall (n=104) Mean (Std. Dev.)	Lead user (n=24) Mean (Std. Dev.)	Non-lead user (n=80) Mean (Std. Dev.)
No suspension	-1.810 (1.303)	-2.128 (1.525)	-1.715 (1.223)
Hard tail suspension	0.739 (1.141)	0.187 (1.261)	0.905 (1.055)***
Full suspension	1.071 (1.583)	1.941 (1.792)***	0.810 (1.425)
Standard transmission	.387 (1.016)	0.471 (1.138)	0.362 (0.982)
Carbon chain guide	0.105 (1.455)	0.256 (1.783)	.060 (1.351)
Gear belt	-.492 (1.273)	-0.727 (1.339)	-.422 (1.253)
Standard wheels & tires	.430 (1.153)	.776 (1.209)*	0.326 (1.122)
Run flat tires	-0.303 (.996)	-0.373 (.697)	-0.281 (1.073)
Fiberglass wheels	-.127 (1.354)	-0.403 (1.381)	-0.045 (1.343)
No power support	1.373 (1.188)	1.026 (1.183)	1.478 (1.176)
Power support	-1.373 (1.188)	-1.026 (1.183)	-1.478 (1.176)
No pedal lock	0.369 (.974)	0.444 (1.028)	0.346 (0.963)
Pedal lock installed	-0.369 (.974)	-0.444 (1.028)	-0.346 (0.963)
Significance: *p<0.1, **p<0.05, ***p<0.01			

Table 38 shows that significant differences in the preference structures of the lead user group and the non-lead user group are present (using a t-test). Thus, this can be interpreted as an indicator that lead user classification was applied successfully.

The evaluation of the suspension differs significantly ($p<0.01$) between both user groups. In addition, the evaluation of wheels and tires shows significant ($p<0.1$) differences on a minor level.

However, this leads to two mountain biking concepts for lead users and non-lead users on the one hand. On the other hand, this shows evidence that MountainCross would be well advised to modify their product line by offering two basic mountain biking concepts: one with full suspension and another one with hard tail suspension. This early result is based on preference measurement and lead user classification solely, but provides valuable information to address future challenges (see in reference to table 31).

The *analysis of internal contributions* (4.3) was made straight forward. Table 39 illustrates exemplary acceptance values for the first 10 users (respondents) and user$_{104}$ and their evaluation of the first eight internal contributions and internal contribution i_{24}. It further provides the resulting mean values and standard deviation for the overall sample and per user group.

The overall rating of *internal contributions* would result in an acceptance of a (i_6) mobile repair set, (i_7) frictionless light concept, (i_{11}) lightweight concept using carbon materials, (i_{12}) comfort seating with reduced weight, (i_{22}) protection foil for the frame, (i_{23}) *dual brake system with intelligent power distribution*, and (i_{24}) retrofitting to use a hub gear. In contrast, the Mann-Whitney-U test revealed significant ($p<0.05$) differences on i_{01}, i_{05}, i_{06}, i_{17}, i_{18}, and i_{19}. It further revealed significant ($p<0.1$) differences on i_{07}, i_{12}, and i_{22} on a minor level. The *evaluation structure between lead users and non-lead users* differs in nine items. Thus, it can be argued that the preference structure of lead users and non-lead users for the (selected) contributions is *similar*. Overall, four internal contributions were found with high ratings by mean value (≥ 0.5) per user group and similar preference structures. i_{06} was added since both user groups provided a positive rating.

Table 39 Acceptance Values for Internal Contributions

Item/User	i_1	i_2	i_3	i_4	i_5	i_6	i_7	i_8	...	i_{24}
u_1	0	0	0	0	0	0	0	1	...	1
u_2	0	1	0	0	0	1	0	0	...	0
u_3	0	0	0	0	0	0	0	0	...	0
u_4	0	1	0	1	0	1	1	1	...	1
u_5	1	0	0	0	1	1	1	0	...	1
u_6	1	1	0	0	1	1	1	1	...	0
u_7	0	0	0	0	0	0	0	0	...	0
u_8	1	1	0	0	0	1	1	1	...	1
u_9	1	0	0	0	0	1	1	0	...	1
u_{10}	0	0	0	0	0	0	0	0	...	1
...	...	...	...	...	...	...	...	...	...	...
u_{104}	1	0	0	0	0	1	1	1	...	0
Lead user (n=24) Mean (Std. Dev.)	0.13 (0.34)	0.38 (0.50)	0.08 (0.28)	0.29 (0.46)	0.04 (0.20)	0.54 (0.51)	0.38 (0.50)	0.33 (0.48)	...	0.50 (0.51)
Non-lead user (n=80) Mean (Std. Dev.)	0.36 (0.48) **	0.41 (0.50)	0.19 (0.40)	0.31 (0.47)	0.31 (0.47) ***	0.77 (0.42) **	0.59 (0.50) *	0.51 0.50)	...	0.62 (0.49)
Overall (n=104) Mean (Std. Dev.)	0.30 (0.46)	0.40 (0.49)	0.17 (0.38)	0.30 (0.46)	0.25 (0.43)	0.72 (0.45)	0.54 (0.50)	0.47 (0.50)	...	0.60 (0.49)
Significance: *$p<0.1$, **$p<0.05$, ***$p<0.01$										

The in-depth analysis of external contributions follows to foster collaborative filtering (4.4). In sum, 80% of the gathered 30 contributions for improvements were shifted to a more specific development level with four innovations being already on the market. The gathered ideas addressed categories of security enhancements, performance improvements, ergonomic solutions, durability, overall design, additional equipment, and preservation. Table 40 illustrates that approximately 63% of the 24 further shifted external contributions addressed the topic of performance improvement. Respondents were allowed to select multiple categories per contribution. Surprisingly, 43% of all external contributions had a focus on a specific problem. This focus is more valuable to the innovation process than problem-focused contributions (see Mahr/Lievens 2012 for a discussion). Problem-focused contributions need to be interpreted (and re-coded) by the project team (see chapter 4.3.3).

Table 40 Selected Contributions by Their Addressed Problem Area

Problem Category	Share in % (n=24)	Selected Examples From the Sample
Safety	27%	Anti-dive braking system
Performance	63%	Adaptive damping system
Ergonomics	43%	Click fix concept for spare parts
Stability	20%	In-frame transmission
Design	13%	Individualised frame concept
Equipment	13%	Pedal extensions
Preservation	7%	Protection foil

However, internal and external contributions partially overlapped. In sum, 18 external contributions resulted from the survey – due to the moderation task.

Collaborative filtering determines the evaluation of external contributions. This started with the development of the similarity matrix of the respondents using Pearson correlation. Table 41 shows an excerpt of the resulting similarity matrix. The excerpt provides an overview for the first 10 users and for the last user u_{104}.

Table 41 Excerpt from the Similarity Matrix

User/ User	u_1	u_2	u_3	u_4	u_5	u_6	u_7	u_8	u_9	u_{10}
u_1	1.00	0.21	0.19	0.18	0.46	-0.37	0.43	0.27	0.44	1.00
u_2	0.21	1.00	0.47	0.26	0.53	-0.04	0.16	0.37	0.31	0.21
u_3	0.19	0.47	1.00	0.11	0.23	0.01	-0.21	0.08	0.18	0.19
u_4	0.18	0.26	0.12	1.00	0.64	0.48	0.53	0.48	0.25	0.18
u_5	0.46	0.53	0.23	0.64	1.00	0.24	0.54	0.57	0.40	0.46
u_6	-0.37	-0.04	0.01	0.48	0.24	1.00	0.09	-0.16	-0.14	-0.37
u_7	0.43	0.15	-0.21	0.53	0.54	0.09	1.00	0.56	0.28	0.43
u_8	0.27	0.37	0.08	0.48	0.57	-0.16	0.56	1.00	0.39	0.27
u_9	0.44	0.31	0.18	0.25	0.40	-0.14	0.28	0.39	1.00	0.44
u_{10}	1.00	0.21	0.19	0.18	0.46	-0.37	0.43	0.27	0.44	1.00
...	...	...	...	...	...	...	...	...	...	...
u_{104}	0.31	0.10	-0.14	0.12	0.36	-0.27	0.56	0.45	0.21	0.31
Interval [-1, 1] (-1= perfect negative correlation; 1= perfect positive correlation)										

Initially, external contributions for analysis resulted in a staged rating matrix from the survey software. Table 42 shows an excerpt for the first 10 and the last external contributions. Derived from this, the evaluation of the binary preferences for external contributions was done using user-based collaborative filtering. The external contribution e_{01} was rated by user u_{32}. This user provided e_{06} to the survey and this contribution was rated by following users (respondents).

The resulting prediction matrix is given below and replaces the original missing values in the evaluation matrix.

Table 42 Excerpt from Evaluation Matrix with CF

		e_{01}	e_{02}	e_{03}	e_{04}	e_{05}	e_{06}	e_{07}	e_{08}	...	e_{18}
Raw data from survey	u_{32}	1	0	0	0	0	-	-	-	...	-
	u_{33}	0	0	1	0	1	1	-	-	...	-
	u_{34}	1	0	1	0	1	0	-	-	...	-
	u_{35}	0	1	1	0	0	0	-	-	...	-
	u_{36}	1	0	0	0	0	0	1	-	...	-
	u_{37}	0	1	1	0	0	0	0	0	...	-
	u_{38}	1	1	0	1	1	1	1	1	...	-
Result from prediction	u_{32}	-	-	-	-	-	0	1	0	...	0
	u_{33}	-	-	-	-	-	-	0	0	...	0
	u_{34}	-	-	-	-	-	-	0	0	...	0
	u_{35}	-	-	-	-	-	-	0	0	...	1
	u_{36}	-	-	-	-	-	-	-	1	...	0
	u_{37}	-	-	-	-	-	-	-	-	...	0
	u_{38}	-	-	-	-	-	-	-	-	...	1

Table 43 presents the resulting mean values that represent the predicted acceptance for all respondents (overall) and per user group.

The Mann-Whitney-U test showed significant differences ($p<0.05$) between the evaluation structure of lead users and non-lead users for external contributions e_5, e_9, e_{17}, and e_{18}. It further revealed significant differences ($p<0.1$) for external contributions e_{06}, e_{08}, e_{14}, e_{16} at a minor level. In sum, eight out of 18 external contributions showed different evaluations by both user groups.

Overall, five internal and three external contributions were rated positively on an overall level and per user group to be possible items for a future product concept. A recommendation was given if the mean value was (≥0.5) and no significant differences were observed between both user groups. This was done by averaging the ratings for all respondents in general. Promising contributions like titan springs for the suspension had to be dropped. However, the project team insisted on incorporating e_{05} – a frame-integrated transmission.

In sum, this led to a variety of modifications to the existing MountainCross concept. Table 44 provides the selected nine contributions that are in the focus for the upcoming lead user workshop. These are only 15% of the overall 59 contributions.

Table 43 Excerpt User-to-Contribution Matrix After CF

		e_{01}	e_{02}	e_{03}	e_{04}	e_{05}	e_{06}	e_{07}	e_{08}	...	e_{18}
	u_1	0	0	0	0	0	0	0	0	...	0
	u_2	1	0	0	0	0	0	0	0	...	0
	u_3	0	0	0	0	0	0	0	0	...	0
	u_4	1	0	1	1	1	1	1	1	...	1
	u_5	1	0	0	0	1	0	0	0	...	0
	u_6	1	1	1	1	1	1	1	1	...	1
	u_7	0	0	0	0	0	0	0	0	...	0
	u_8	1	1	1	1	1	1	1	1	...	1
Merged	u_9	1	0	1	1	1	0	0	1	...	1
data for	u_{10}	0	0	0	0	0	0	0	0	...	0
further	...	...	...	...	...	...	...	...	...	...	...
analysis	u_{32}	1	0	0	0	0	0	1	0	...	0
	u_{33}	0	0	1	0	1	1	0	0	...	0
	u_{34}	1	0	1	0	1	0	0	0	...	0
	u_{35}	0	1	1	0	0	0	0	0	...	1
	u_{36}	1	0	0	0	0	0	1	1	...	0
	u_{37}	0	1	1	0	0	0	0	0	...	0
	u_{38}	1	1	0	1	1	1	1	1	...	1
	...	...	...	...	...	...	...	...	...	...	...
	u_{104}	0	1	1	0	1	0	0	0	...	1

Lead user (n=24) Mean (Std. Dev.)	0.71 (0.46)	0.37 (0.50)	0.50 (0.51)	0.42 (0.50)	0.38 (0.50)	0.25 (0.44)	0.42 (0.50)	0.29 (0.46)	...	0.21 (0.42)
Non-lead user (n=80) Mean (Std. Dev.)	0.56 (0.50)	0.39 (0.49)	0.52 (0.50)	0.53 (0.50)	0.62 (0.49) **	0.47 (0.51) *	0.47 (0.50)	0.51 (0.50) *	...	0.51 (0.50) ***
Overall (n=104) Mean (Std. Dev.)	0.60 (0.49)	0.38 (0.49)	0.52 (0.50)	0.49 (0.50)	0.56 (0.50)	0.42 (0.50)	0.46 (0.50)	0.46 (0.50)	...	0.44 (0.50)
Binary measurement [0; 1] (0= not preferred; 1= preferred)
Significance: *p<0.1, **p<0.05, ***p<0.01

Table 44 Additional Product Components after Preference Measurement

Additional Internal Contributions	Add. External Contributions
i_{06}: mobile repair set	e_{01}: extended barends
i_{11}: lightweight concept using carbon	e_{03}: advanced pneumatic suspension
i_{12}: comfort seating with reduced weight	e_{05}: frame-integrated transmission
i_{23}: dual (smart) brake system	e_{15}: advanced hub gear
i_{24}: retrofitting to use a hub gear	

4.4.6. Phase 5 – Lead User Workshop

The *pre-assessment of internal and external contributions* in relation to available resources, feasibility, and production (5.1) follows the survey and preference measurement. This was done internally by the project team. Basic questions were for example: "How can we adapt a new concept to fit our production?", "What was the basic problem of the contribution?", "How can we improve the suggested solution to fit our standard product?" among others.

Practice has shown multiple approaches to *perform an evaluation task*, like the function point method and the definition of storytelling points in software development, or QFD and target costing in NPD. Underlying scoring models, risk analysis, value trade-offs, and rankings offer a great variety and thus this remains individualised for each application field. Here, the criteria were chosen in reference to the master project plan (see table 32) and a previous analysis (see table 31). Each contribution was evaluated in terms of *"production feasibility"*, *"re-use of current components"*, *"expected projection costs"*, *"relation to existing partners"* etc., and *"existing IP protection"*. The evaluation itself was the result of an argumentation within the – interdisciplinary – project team. The relation to existing partners was a criterion of the senior management. In sum, this rating will be a general anchor to estimate the feasibility with existing and future resources.

The *concern about present patents* was a major result of this preparation. For example, the enterprise Pinion holds a patent for a frame integrated transmission (external contribution e_5, see for example DE 102010051727 and DE 102011106107). This is valuable information since patent litigations are major threats for a SME. Today, another competitor has filed an opposition to the patent (on July 17th 2013 in reference to the process of sensing torque) and this confirms previous concerns. Additionally, it offers further possible partners that were not identified during market exploration and can be invited to the workshop. In addition, a dual brake system with intelligent power distribution (i_{23}) would make use of a technique that is already known (see patent DE 203093755 that expired on Jan 1, 2010). Overall, multiple foreign patents and utility models were found and thus one major task of the workshop would be to find appropriate workarounds with lead users.

Overall, important *background information* is available for the workshop team and the moderator of the workshop at this stage including: (1) a basic mountain biking concept for lead users and non-lead users, (2) ratings for internal contributions and (motivated) in-house engineers, (3) predictions for external contributions including lead users, (4) an internal assessment of effort for further development, and (5) information about patent implications.

Administrative tasks (5.2) followed to organise the workshop and to invite internal and external contributors to participate. This included for example the generation of non-disclosure agreements (NDA) for each participant. All identified lead users were invited, but only eight were willing to join the workshop. The major goal of the applied 2-day workshop was *technical specification, concept generation*, and *concept refinement* (5.3). The workshop itself was performed in orientation to Churchill et al. (2009) and resulted in one final mountain biking concept.

The previous assessment of contributions was used to *identify major topics of interest*, like workarounds for contributions with existing patents and adaptive tasks to fit the given resources of MountainCross. The *resulting task from preference measurement* was a redesign of the market available frame to include a locked rear suspension within a carbon fibre frame. The major problem may result from rapid forces working on the hard frame. However, one of the greatest advances of a carbon aramid alloy is the possibility to design frames and not to rely on a traditional diamond design. Frame design was one of MountainCross' strengths. The dual brake system (i_{23}) with an intelligent distribution to prevent accidents in highly dangerous situations was a major challenge. Although, an anti-braking lock system for bicycles is well-known, it needs major refinement to fit MountinCross' target group and to work around current patents. Contributions e_{01} (extended barends) and i_{06} (mobile repair set) were an easy win for MountainCross and could be managed by adapting the on-site purchase process to allow individualisation. The external contribution e_{03} (advanced pneumatic suspension) and e_{05} (frame-integrated transmission) were defined within the workshop and is now developed with former partners and suppliers. Internal contribution i_{24} (hub gear) was cancelled. The resulting mountain bike makes use of a carbon fibre frame, standard full suspension with an advanced pneumatic system and lockout function, a frame-integrated transmission, a smart brake system, and a frictionless light system with passive illumination.

The *re-test of external contributions* (5.4) is designed as an expert interview with a 5-point Likert scale. Overall, 23 bicycle professionals and sales agents were interviewed in multiple German cities like Berlin, Cottbus, and Dresden to assess the novelty and the market potential of the contributions.

Table 45 Evaluation of Novelty and Market Potential by Experts

Provided Lead User Contributions[1]	Chi-Square Test Within Experts (n=23)		Unknown by # Experts
	Novelty Mean (Std. Dev.)	Market potential Mean (Std. Dev.)	
e_{01} extended barends	2.05 (1.17)**	2.27 (0.94)*	0
e_{02} click fixation	2.50 (1.23)**	2.55 (1.10)	1
e_{03} advanced pneumatic suspension	2.55 (1.71)***	*3.64 (1.18)*	1
e_{04} adjustable seatpost	2.26 (1.24)	3.52 (1.12)	0
e_{05} frame-integrated transmission	2.50 (1.63)*	2.86 (1.28)	1
e_{06} perforated seating	2.09 (1.24)**	2.26 (0.92)	0
e_{07} titanium springs for suspension	1.81 (0.873)**	2.19 (0.928)	2
e_{08} anti-dive braking system	2.59 (1.06)	2.35 (1.12)	6
e_{09} carbon drive shaft	2.09 (1.27)**	1.86 (0.94)*	1
e_{10} flat-pedal with concealed mechanics	*3.11 (1.50)*	*3.00 (1.05)***	4
e_{11} individualised handlebar	1.61 (0.78)*	2.70 (1.15)	0
e_{12} individualised pedal extensions	2.10 (1.18)*	2.81 (0.93)**	2
e_{13} individual frame concept	2.05 (1.07)**	2.86 (1.06)	0
e_{14} innovative chain retention	2.40 (1.14)	2.45 (0.89)*	3
e_{15} hub gear	2.83 (1.50)	2.39 (1.20)	0
e_{16} adaptive seat	2.50 (1.54)	*3.14 (1.39)*	1
e_{17} winter-proved mud flap	2.71 (1.55)	2.90 (1.18)	2
e_{18} protection foil	2.04 (1.26)*	*3.00 (1.13)*	0
(1: respondents were provided with a more detailed description of the contribution presented in the survey); (1= already known/no expected market potential; 5= strong novelty/high expected market potential); Significance: *p<0.1, **p<0.05, ***p<0.01; 5-point-rating scale			

Table 45 shows the results and indicates that heterogeneity existed in ratings of novelty and market potential. In general, experts have shown great interest in the presented user contributions. Especially, e_{03} (advanced pneumatic suspension) stimulated discussion and was in doubt of being applicable, but was rated with high potential. The major components of an advanced pneumatic system, a smart braking system, and a frame-integrated transmission received homogeneous ratings in market potential. This partially confirms previous recommendations by collaborative filtering. External contribution e_{03} shows strong significant (p<0.01) differences by expert ratings on novelty, but was widely accepted as possessing a high market potential. Further, experts agreed on a high market potential for e_{05}.

The main purpose of this re-test was to *confirm predictions* on external contributions. Overall, 66% (two out of three external contributions, except e_{01} which was rejected and e_{05} which was included in the workshop but rejected by collaborative filtering) of the recommendations by collaborative filtering were confirmed by expert interviews in market potential.

The distribution to the *senior management* finalised the lead user project (5.4). The senior management is now in charge to set go for further product development.

A possibly developed and manufactured prototype of the final mountain bike may be presented to MountainCross' handpicked customers to discover how they will react to the new features and how they might change their usage behaviour. An additional simulated test market may show convincing results to introduce the product to the market for a reasonable price of about EUR 800, which is settled in the middle class price segment. The innovation goal of an expected revenue of EUR 76 Million in the fifth year can then be achieved by selling more than 95,000 MountainCross bicycles. This would equal a market share of 24% in mountain biking and 2.4% market share in general cycling in Germany. In comparison, the real-existing competitor MIFA (Mitteldeutsche Fahrradwerke AG) with its acquired mountain bike brand Steppenwolf sold 546,000 units in 2012 (MIFA 2013).

In general, the underlying empirical data for this application at MountainCross is derived from previous preference measurement (see Sänn/Baier 2014). Collaborative filtering was applied on the present data to provide this application example. The expert interview was conducted with the help of undergraduate assistance and confirmed the predicted values.

4.5. Implications

The proposed Preference-Driven Lead User Method aims at lowering the risk of a failure when the lead user method is applied. Thus, it is expected that the new method can improve the effectiveness of NPD and lead to multiple advantages, especially for SMEs. It is expected to lead to a decreased novelty but an increased market potential of a new product concept under development.

In detail, the application of the Preference-Driven Lead User Method is assumed to address challenges that were identified in chapters 2 and 3. Among these *challenges* are a strong influence of the functional fixedness next to the NIH and an overall influence of the results by relying on lead users solely, like addressing extreme needs that ordinary customers may never have.

Thus, a product concept from a lead user workshop may result in a long-term diffusion process or may fail in diffusion entirely. This is also the case when the product concept fails to convince senior management. A failure of the traditional method leads to expensive re-work and increases *uncertainty* at senior management's level. A further background of restricted resources demands different solutions. Several authors have claimed that methods of open innovation, like the lead user method, need to be updated for an improved usage in SMEs.

This solution is proposed as the Preference-Driven Lead User Method that addresses the research question *"Can the lead user method and preference measurement be combined to result in an integrated method for new product development?"*. The new method is expected to benefit from preference measurement and collaborative filtering to prepare the workshop setting in general and to fill the gap between 'ideation and evaluation' and concept development. Methodological adaptions are focused on revealing similar evaluation structures between a group of lead users and non-lead users. This covers preference measurement for state of the art attributes and a binary evaluation of internal and external contributions that is supported by collaborative filtering. The proposed method is exemplary applied in the case of the fictive company MountainCross. The application example fosters on actual preference values and applies collaborative filtering to illustrate the basic methodology. The conducted expert interview confirms previous assumptions of an increased market potential.

5. Empirical Comparison in the Field of Industrial IT Security

5.1. Overview

The Preference-Driven Lead User Method is applied within the field of *industrial IT security* for Critical Infrastructures (CRITIS). This field has become of major importance in recent years since IT security vulnerabilities started to become prominent in the media. Today, relevant trends like the Internet-of-Things and Cyber-Physical Systems require reliable IT security mechanisms to work properly. In contrast, prominent cyber-attacks such as *Stuxnet* revealed the vulnerability of these systems that cannot be protected by simply adopting traditional IT security mechanisms. Industrial IT systems deal with low performance devices and thus any modern IT security mechanism would result in excessive processing and energy effort for the device. This innovation environment frames the application field.

The traditional lead user method was employed in the project *Enhanced Security for Critical Infrastructures*. The research goal was to enable a multi-level security system that is able to use state of the art IT security algorithms within embedded components. The application of the new method was a follow-up work after completing the traditional method. The comparison of the Preference-Driven Lead User Method and the traditional lead user method is fostered on a rating of the outcomes of the applied methods in terms of market potential and novelty.

Chapter 5.2 introduces the broad field of Industrial IT security. This provides an overview of the general challenges in this application field and deepens the understanding of IT security in CRITIS. It further discusses user innovations in this field and refers to early experiences with a heterogeneity of needs and a suprising homogeneity of evaluations of novel security-related functions. Chapter 5.3 describes the application of the traditional lead user method within the before mentioned joint project of IHP GmbH and Brandenburg University of Cottbus in detail. It ends with a brief overview of the resulting concept. Chapter 5.4 provides an overview to the application of the Preference-Driven Lead User Method. In phase 5 of the new approach, both resulting concepts were tested using simple ratings. Respondents that participated either in the traditional method or in the proposed method were asked to evaluate the estimated market potential and novelty. Chapter 5.5 summarises the application and provides further implications.

5.2.   Introduction to the Application Field

5.2.1. The Field of Industrial IT Security

Over the past years, the field of industrial IT security (IIT security) has been prominent in the news. In general, IT security describes a field of *preventive and reactive mechanisms to protect security properties*. These properties are defined as confidentiality, integrity, and availability of information (see norm ISO/IEC 27000:2009). Various research include additional security goals such as authenticity, accountability, and non-repudiation (see in reference Eckert 2008).

The basic concepts of *security and safety* are to be distinguished and are often confused in practice. Safety describes that a specified functionality is coherent with the actual functionality of an IT system. This is not in the focus of security. In contrast, security describes that an IT system can only obtain a specified status that does not allow an unauthorised data manipulation or information loss (see Eckert 2008). For example, driverless transportation systems (DTS; see e.g. Meinberg 1989 for an early description of their functionality) need to be safe – they operate within their pre-defined areas. The communication towards multiple DTS needs to be secured in order to maintain functionality in accordance with required safety levels.

IIT security became one of the major topics in the ICT world since *Cyber-Physical Systems* (see Sandberg et al. 2015, CPS) became more and more important. CPS are *embedded systems* that collect, process and forward data to other embedded systems and central processing units. CPS monitor real-world phenomena and control real-world systems. This connects the digital world to the real world and is highly associated with the IoT. Unlike dedicated hardware (e.g. in industrial automation systems), these systems were never meant to communicate together (e.g. sensors in washing machines) and this raises concerns about IT security. However, benefits of inter-connected systems like an optimised resource management are the reason why the usage of embedded systems, in e.g. power plants and production sites, is constantly growing. Today, embedded systems are the hidden champions of the German industry with a domestic market volume of about EUR 20 billion. The global market of embedded systems can hardly be quantified, but studies assume that about 98% of the global chip manufacturing industry is related to this field. Overall, more than 4 billion embedded devices are sold per year (cf. BITKOM 2008, BITKOM 2010).

That is why the IoT and associated trends like *Industry 4.0* are of great relevance for semiconductors and chip designers (see for example financial analyst reports like Blankenhorn 2015). Figure 10 illustrates such an embedded device - the trusted sensor node of IHP GmbH. This node can be integrated into multiple application fields to provide basic aspects of IT security like secure and encrypted data exchange for distances up to 500 meters. *Supervisory Control and Data Acquisition systems* (SCADA) aggregate data from such embedded devices/systems and trigger specific control functions, e.g. to regulate the water level at a dam. This node would perform and secure the data transmission in such a case.

Figure 10 Trusted Sensor Node by IHP GmbH

Today, IIT faces major changes since wired, proprietary, and individualised communication systems and protocols are now substituted by e.g. *wireless solutions* that deal with common vulnerabilities (see Stecklina/Sänn 2012).

A fundamental problem for IT security is the Internet itself. It was not designed for today's security and flexibility requirements (see Stuckmann/Zimmermann 2009 for a brief discussion). Worse, the Internet was never designed to deal with *untrustworthy users.* New classes of applications and devices, modern operational and management requirements, and a continuously patching of protocols to fit actual requirements generate further challenges. Unfortunately, this generates new challenges for IIT security that cannot be solved with traditional mechanisms and demand radical innovations (see ESCI 2011).

The 1st reason to explain this demand can be found in *limited hardware resources.* IIT systems deal only with a fraction of standard IT systems' hardware performance and thus familiar state of the art IT security solutions cannot be adopted. This is triggered by e.g. long-range update cycles of the systems.

The 2nd reason is that these standardised systems and protocols were never designed to operate in an environment with such a *high integration level* (see Sänn/Krimmling 2014 for a short discussion and survey results for future IIT security directions). The level of system integration describes the connection of different components in an industrial environment and their complexity.

The 3rd reason is an *increased requirement for flexibility* that is the major driver of such complexity. SCADA systems are now connected to wide area networks – the Internet – and provide remote services for e.g. maintenance reasons. This is why they are facing serious security challenges that lead to a multilayer problem (in reference to the OSI model). This multilayer problem describes the fact that multiple layers in the network are now exposed to the public and face serious threats because of an Internet connection.

The 4th reason can be seen in a *heterogeneous market structure* with many proprietary communication protocols (e.g. Ethercat, Profinet, Profibus, and Modbus). The heterogeneous structure leads to use of protocol interfaces and protocol converters to handle *cross-protocol communication* to realise modern functions in automation systems (see e.g. Klebe-Klingemann/Schneider 2010 for this scenario at the Eon Thüringer Energie AG). Industrial networks typically consist of human machine interfaces (HMI), fieldbus controller (PLC), and field devices (sensors and actuators) that are connected via fieldbus. Fieldbus coupler connect various devices (analogue and digital) to the fieldbus. Figure 11 illustrates such a complex topology that needs to be addressed to fulfil future demands in IIT security (see Sänn/Krimmling 2014).

Thus, SCADA systems themselves became a *relevant point of failure* in an actual industrial communication network and are responsible for significant affects to availability, integrity, and confidentiality, e.g. by denial of service attacks (DoS) or SQL injection. Larkin et al. (2014) provide an alarming evaluation of security solutions for SCADA devices.

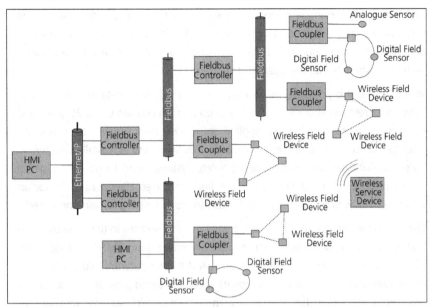

Figure 11 A Complex Topology from an Industrial Network
(Reference: Illustration in Reference to ESCI 2011)

A possible solution to enable SCADA security would be using a preventive and reactive part of IT security. Preventive security spans *encryption* and *authentication* algorithms to secure a data connection from being attacked. Reactive security employs *Intrusion Prevention and Intrusion Detection Systems* (IPS/IDS, see for example Pathan 2014 for a brief summary of state of the art security solutions for automation systems by Krimmling/Langendörfer, pp. 431-448) and further security policies to regulate the actual traffic and to detect malicious behaviour, e.g. by anomaly detection or signatures in the traffic. Both parts are necessary to guarantee a system's functionality, e.g. monitoring real world phenomenon in CPS.

The 5th reason describes the *principles of the security market in general*. Traditionally, IT security works partially by *patching disclosed* (software) *vulnerabilities* that occurred while analysing protocols, traffic handling in software, and memory allocation etc. and thus software vendors need to keep up with actual zero-day vulnerabilities. Unfortunately, this fails from time to time in traditional IT environments. The case of Oracle's Java Runtime Environment in 2013 illustrated that. Java faced multiple vulnerabilities that affected many embedded devices, too, and allowed for example remote code execution (see e.g. the company's own software security assurance blog Maurice 2013 for a broad description of its security status and fixing policy). This scenario is also applicable for firmware on embedded devices that are illustrated in figure 11.

In contrast to this principle, IIT systems are designed for long-term usage and to serve in extreme environments. Technical update periods are up to 25 years and every new patch has to be applied in a *test environment* to check for interoperability with other embedded devices and further installed software. Rarely, update procedures fail and complete systems need to go offline which makes prior tests of patches even more important (see Eikenberg 2015 for an example). This is time consuming and thus various IIT systems remain unpatched.

Overall, all five reasons require a new *awareness of IT security* in this industrial field and demand new solutions (see for example Krimmling/Sänn 2015 for a description of traditional centralised security solutions and a new decentralised approach in IIT). In addition, non-technical constraints like a shortened time of IT projects and less profit influence the overall *quality of employed security mechanisms* (Elger/Haußner 2010). This is may link to misconfiguration of the IT systems and missing knowledge about the specific IT architecture and security strategy. The standard example is an unknown, but a physical or virtual existing network link connecting the traditional "office" LAN and the SCADA network.

5.2.2. IT Security in Critical Infrastructures

The restrictions mentioned above are immanent for *Critical Infrastructures* (CRITIS). CRITIS are defined as facilities that will influence the social community in a significant way when they fail (see BMI 2009 for the definition and e.g. Elsberg 2012 for a fictive scenario of a cyber-attack on CRITIS in Europe). Figure 12 provides an aggregated overview to CRITIS sectors in accordance with the classification from the German Federal Ministry of the Interior (BMI). The U.S. Department of Homeland Security mainly accepts this classification, but also highlighted commercial facilities, *critical manufacturing sites*, dams, the defence industry, and nuclear reactors as a major sector.

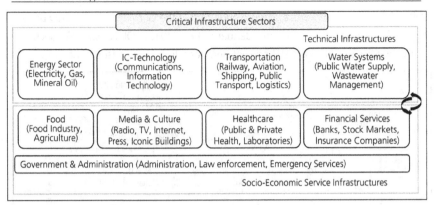

Figure 12 Sectors of Critical Infrastructures in Germany
(Reference: Illustration in Reference to BMI 2009)

Securing CRITIS is a complex task since they are partially in private hand, but serve the public domain. They are further *integrated* and *interdependent*. This leads to a chain of authorities and companies that are concerned about IT security for CRITIS.

August/Tunca (2011) argue that "Internet security is clearly in the public's best interest, but, thus far, private companies and end users did not necessarily have adequate economic incentives to incur costly investments to create more secure software products and properly maintained computing systems" (August/Tunca 2011, p. 935). This leads to a general dilemma of understanding IT security in CRITIS and the mentioned responsibility. The authors compare the effectiveness of three *software liability policies* to prove their point and found mixed results. Possible scenarios were (1) vendor liability for damages, (2) vendor liability for patching costs, and (3) government imposed security standards. However, the latter alternative was for example applied by the German Federal Office for Information Security – the *IT baseline protection policy* – and the whitepaper "Requirements for Secure Control and Telecommunication Systems" by the German Association of Energy and Water Industries (BDEW).

Besides all regulations and countermeasures, prominent cyber-attacks like *Stuxnet* (see Langner 2011 and Fidler 2011), Flame, Duqu etc. found their way to CRITIS, caused severe damage, and were featured in the news. In addition, recent IT security breaches at Lockheed Martin, Sony, and BMW showed that this is still an issue (see e.g. Spaar 2015). Sänn/Stecklina (2013) provided an overview of CRITIS related IT security incidents that were published at the German heise security portal from September 2nd 2011 to January 17th 2013. Obviously, IIT security vulnerabilities in CRITIS are still present and need to be solved (see figure 13).

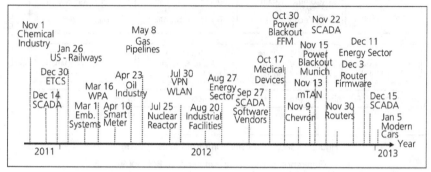

Figure 13 Selected Incidents from September 2nd 2011 to January 17th 2013

(Reference: Illustration in Reference to Sänn/Stecklina 2013)

An approach to address the vulnerabilities mentioned is given in figure 14. This illustrates the multi-layer approach that was in the focus of the project Enhanced Security for Critical Infrastructures (ESCI). The project was hosted at IHP GmbH, made use of the lead user method and serves as the application field. Overall, security for IIT is a major topic that is highly related to modern industrial espionage and therefore an effective solution is needed.

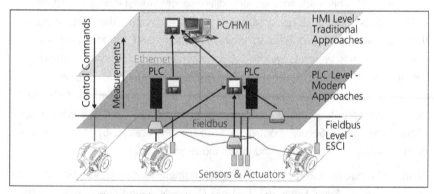

Figure 14 Distributed and Reactive Configuration of ESCI

In sum, IIT security is dealing with complex problems. These demand radical changes and breakthrough solutions to take the leap from traditional bug fixing and patching to be ahead of hackers and professional cyber attackers.

5.2.3. User Innovations in IT Security

The "hacking" term describes an interaction with a technology-based system in a playful and explorative way (see Paradiso et al. 2008). Thus, this includes inventors and describes everyone who customises a technology-infected system. The motivation to hack is not necessarily e.g. revenge, mayhem or digital vandalism, but is also related to the fundamental motivation of joy, learning and being part of a community (in reference to subchapter 2.2.5 and Lakhani/Wolf 2005).

Franke/von Hippel (2003) made experiences with the lead user construct in the general field of IT security and observed *user innovations*. Lead users were writing own code to extend the Apache web server software with plugins to address multiple security features. Users of the software were not satisfied with provided standard security functions and thus innovated to fit their needs. This was possible since the web server software is signed as open source software.

The most important aspect is the *heterogeneity of needs* observed in the sample. The authors started with a list of 45 security-related functions for the Apache web server that was generated by using experts from this field. Overall, almost 50% of their respondents added *new security-related functions*. In the end, 137 distinct security-related functions were identified as user needs. Interestingly, heterogeneity was equal across all respondents. Franke/von Hippel (2003) pointed to the increasing complexity of the web server that triggered new needs in IT security. Further, no significant differences in the judgement of the importance of the security-related functions could be observed between the respondent groups, although respondents with a high skill-level provided more new contributions. This situation from late 2000 is *similar to the situation of today's IIT security*. The first version of the web server from 1996 provided only a fraction of today's functionality and was relatively simple. This is comparable with today's SCADA systems' security and programming software. As a consequence, online discussion boards evolved and serve as a source for user innovators who provide various source code, e.g. for communication libraries, to administrators of SCADA systems. Today, hacking also offers a new way for doing business. This serves as an indicator that there is a variety of relevant ideas to secure and unsecure IIT systems, which need to be found (see for example Gambardella/Giarratana 2007 and Ghosh/McGraw 2012 for further outlines).

5.3. Application of the Traditional Lead User Method

5.3.1. Preface - Enhanced Security for Critical Infrastructures

The application of the lead user method was embedded in the project of *Enhanced Security for Critical Infrastructures* by IHP GmbH. The project was funded by the German Federal Ministry of Education and Research (BMBF) under the grant number 03FO3102. The general application field was pre-defined by the upper management and had a focus on Cyber-Physical Systems (CPS). The related technical context of this project is the field of IIT security and is presented in detail by e.g. Sänn (2011) and Wessling (2011). The project team became a member of the CAST e.V. during the project and was an active member of the CRITIS research group of the German Informatics Society (Gesellschaft für Informatik e.V. – GI).

The project further organised and moderated a dedicated workshop that was hosted at the Fraunhofer IGD in Darmstadt (see http://www.cast-forum.de/workshops/infos/165) with leading IT security experts from several application fields (see e.g. Borchers/Wilkens 2012 for a coverage by heise security and Herbst 2012 for an interview at deutschlandfunk with the ESCI project coordinator). The ESCI project ended on March 31st 2013. The result of the project was presented at the Hannover Fair 2013 (see Krimmling/Sänn 2015 for it functionality) and was awarded with the research transfer price 2013 of the Brandenburg University of Technology Cottbus.

The research goal was to enable a *multi-level security system* that is able to use state of the art IT security algorithms within embedded components. A major consensus was the development of a *distributed and reactive security platform* (see figure 14). Authentication, authorisation, and a secure exchange of cryptographic keys were supposed to be spanned over multiple levels to realise a secure communication between system components. Prevention is supposed to realise trusted communication relationships and to provide notifications of the security status to network components. The reaction is characterised by an *intrusion detection system* and further mechanisms to engage countermeasures that requires new approaches. This includes for example the development of an *Elliptic Curve Cryptography* (ECC) implementation for usage in low-performance devices.

5.3.2. Phase 1 – Preparation for Project Launch

The application of the traditional lead user method started in October 2009. The project aimed at (1) a technological development based on the given *project proposal*, (2) an outline of the state of the art in the application field, (3) an analysis of future economic implications, and (4) a proof of technical applicability for solving relevant IT vulnerabilities. The application of the lead user method was chosen to support the technological development by incorporating market feedback and preliminary lead user solutions. The joint project team consisted of five scholars in computer science and one scholar in business informatics. The team of six scholars was led by three Professors from computer science and economics.

Team foundation (1.1 in reference to table 28) was in the focus of the first few meetings and started with an introduction of all team members. The technological vision was a *peer-to-peer intrusion detection and prevention system* (P2P-IDS) for wireless sensor and actuator networks (WSAN) in CPS. The in-depth analysis of the vision was discussed in the following meetings and revealed research challenges to the team. Figure 15 presents the vision of a P2P-IDS overlay. The cooperation overlay would negotiate cooperative tasks to perform and to schedule analyses. The information overlay would scan for available processing resources. The network infrastructure provides the traffic for an analysis (see König 2005).

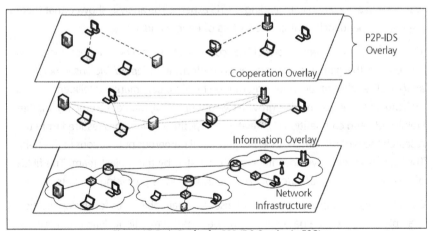

Figure 15 Early Draft of a P2P-IDS Overlay in ESCI

The main task of the *project schedule* (1.2) was to generate a common understanding of the research field and to establish a common technological vision and problem awareness. In general, WSANs are in need of security, but provide only *limited hardware resources* (see the in-depth descriptions in chapter 5.1). Thus, state of the art security operations are very costly in terms of energy consumption and CPU load (see for example Portilla et al. 2010 and Krimmling/Langendörfer 2014 for basic challenges). This emphasised technological constraints, like applicable methods for traffic handling and *latency*, and draws additional requirements, like *real-time availability* in complex system environments.

Overall, the project team identified multiple challenges to implement the vision of a P2P-IDS overlay, among them were: (1) latency between multiple networks, (2) handling of excessive traffic, (3) limited processing resources, (4) efficient search for new network nodes, (5) validation of analytic results, (6) authentication of sensor nodes, (7) encryption, (8) hardware and energy resources, and (9) privacy.

The economic environment was given by the project proposal and put Critical Infrastructure Protection (CIP) in the focus. However, the field of CRITIS offers various application scenarios. Thus, the project team collected information about relevant CRITIS sectors and identified superficial trends and technological streams. The further analysis concentrated on *interdependencies among CRITIS sectors*. This was done in association with leading CRITIS experts. The analysis showed that there are main dependencies within the sectors of energy and ICT.

The economic perspective on CRITIS concentrated on key business facts and aggregated the gathered information to perform an *adapted McKinsey portfolio analysis*. The relevant facts were interpreted by secondary market intelligence using indicators of expected market growth (weight factor 4), market volume (2), market stability (3), and competition (1). Relative competitive advantages resulted from the research focus of IHP GmbH, its experience, and numerous publications in this field. The relative competitive advantage was evaluated by the project team. The fields of *energy and industrial automation* were evaluated as promising markets.

Overall, the master project plan summarises the main objectives for this application example and builds the basement for phase 2 of the traditional lead user method. In reason of the structure of the research programme *ForMaT* by the German

Federal Ministry of Education and Research (BMBF) the lead user project had to be separated in two parts.

Table 46 Master Project Plan for ESCI

Scope	Definition of the Scope
Product category	P2P-IDS overlay for CPS
Target markets	Business-to-business surrounding; CRITIS
Applications of interest	Energy automation, industrial automation
Desired project outcomes	Solutions for identified research questions (auth., coop., security), prototype, pilot application, spin-off
Key business goals & constraints	2nd round funding, 6 months time constraint

The 1st part focused on project grounding, identification of key trends and customer needs, as well as lead user identification (up to operational step 3.1 in table 28). The 2nd part focused on an exploration of lead user needs (validation of 3.1 and accomplishment of 3.2 and 3.3) and the improvement of solution concepts with lead users and experts (phase 4). The 1st part of the project ended in March 2010. The 2nd part was dependent on a granted project funding and started in March 2011.

5.3.3. Phase 2 – Identification of Key Trends

The McKinsey portfolio analysis confirmed the relevance of energy and industrial automation. Interestingly, *industrial automation is described as a cross-sectional application field for CRITIS and non-CRITIS* that includes energy automation, smart home (facility) automation, and process automation. This is verified by the actual *convergence of both markets* and results in commonly used SCADA technologies. Thus, energy and industrial automation were seen as analogue markets.

The project team started to collect further information about relevant CRITIS sectors and intensively screened the market for trends and technological streams (2.1; see Baier/Sänn 2013, p. 804). These were gathered by analysing academic theses, journal articles (relevant journals for the application field in general and for IT security in particular), and business magazines as well as online reports (discussion boards, blogs etc.). Subsequently, primary research was conducted in order to add to this secondary research. *In-depth interviews* (face-to-face) with regional experts in the field of energy and industrial automation as well as telephone interviews with lobby associations (e.g. BDEW, BITKOM) were employed. Since this was a joint project with Brandenburg University of Technology Cottbus, its chairs for energy automation and industrial automation offered beneficial local contact to validate trends and to discover new problem fields. Identified trends were distinguished to cover social, economic and technical streams among them: (1) expansion of renewable energy, (2) unbundling of the German energy sector, and (3) convergence with industrial automation systems, in general. Table 47 provides an overview of relevant trends that were collected by the project team.

Table 47 Selected Identified Trends from Trend Analysis in 2009 and 2010

Selected Trend Category	Selected Identified Trends
Social and political trends	Expansion of renewable energy and EEG
	Increasing awareness by federal ministries and institutions
	Increasing awareness of IT security and espionage
	Increasing sensibility in reason of negative examples from North America (power failures, attacks on CRITIS)
	Bring your own device and the internet of things
	Lükex scenario and wikileaks
Economic trends	Unbundling of the German energy sector/system
	Increasing wireless connectivity for cost reduction
	Financial crisis and recession
	Decentralised power generation and smart metering
	Business models for own data centres
Technical streams	Convergence with industrial automation systems
	Termination of ISDN phone lines from Deutsche Telekom
	Protocol IEC 60870-5-104 and smart grid technologies
	Standardisation of common information model (CIM)
	Implementation of industrial Ethernet components and protocols
	Wireless connectivity for remote stations in energy networks
	Increasing data connectivity within SCADA systems
	Increasing remote functionality of SCADA systems

The *evaluation of trends* was further fostered on interviews with experts and lecturers from the application fields and discussed implications of trends for industrial automation in general. Relevant trends that were employed to support lead user classification were e.g. (1) business models for own data centres (and the knowledge about IT security implications), (2) investments in decentralised power generation, and (3) current development of IT security concepts for own automation systems (see Baier/Sänn 2015 for a brief description of trend analysis and Sänn/Krimmling 2014 for actual trends in the field of IIT security that still remained in the application field).

The *data collection plan* would guide the project team through the interview process and was separated into three parts. The 1st part asked about IT security in general. This included questions to cover IT security requirements, IT security awareness, and hardware implications. The 2nd part asked about the identified social and economic trends. The 3rd part asked about known development streams in analogous application fields and for referrals.

Lead user classification (2.2) used the original definition by von Hippel (1986). This was extended by aspects of lead userness, among them: (1) an own development, (2) significant R&D expenses in IT security, (3) implementation of industrial Ethernet components, (4) being directly affected by termination of ISDN phone lines, (5) pioneer characteristics for developers and manufacturers of automation hardware and software, and (6) early-adopter characteristics and reputation. The selection of indicators for lead userness were discussed with a group of experts that pointed to the relevance of own investments in projects for distributed energy generation and an effort for developing own IT security approaches (see Baier/Sänn 2015).

The *observation of needs* (2.3) was done by site visits, telephone interviews etc. – thus, the project team gained insights into the application field. This included site visits in a power generation plan at Vattenfall Europe and in an energy distribution system at Stadtwerke Cottbus in January and February 2010.

5.3.4. Phase 3 – Exploration

The relevant application fields were assessed for *lead user identification* (3.1) using the pyramiding approach. Already present contacts of the team members built the basis and were extended by interviews within the Institute of Power Engineering at BTU Cottbus and previous experts. Regional contacts started the pyramiding process. In reference to Churchill et al. (2009), the data collection plan was used to guide the interviews and to continuously learn about the market.

The contacts confirmed the *relevance of the research topic* and expressed interest to support the project team. In addition, contacts were made to national companies such as Stadtwerke Jena-Pößneck and Envia Networks. This reflected the supply chain of the energy market, too. The field of industrial automation was equally addressed. Overall, n=162 contacts were interviewed with an average of about three in-house referrals and an average of 0.43 referrals per interview (overall referrals per interview is about 0.96 according to the eight examined lead user studies by Poetz/Prügl 2010). When respondents provided a referral, then the average amount of referrals was 2.1. A maximum amount of 10 referrals was given by two respondents each – a full-time research associate and a regional IT security expert. It was observed that the amount of referrals decreased with the progress.

The promising contacts were asked to take part in an online-survey to confirm their lead user status (see Baier/Sänn 2015). This verified the gathered trends, too, and gave a brief *overview of the respondents' demands* to secure specific security goals, like availability (see Sänn/Krimmling 2014 for an updated version of the survey).

In sum, 10 lead users with a business-to-business background were identified in the application field of energy automation and were willing to *collaborate within the project*. At this point, the lead user method had to stop since the 1ˢᵗ stage of the BMBF-ForMaT research programme ended.

In March 2011 the project continued to last until March 2013. Since April 2010 Stuxnet was a major topic within the field of IIT security. Stuxnet addressed industrial Ethernet components of energy automation systems and *proved the identified trends and vulnerabilities*. Further, the Fukushima disaster influenced the identified trends and was responsible for a dramatically increased awareness for plant security and safety. Stuxnet and Fukushima influenced the business perspective by shifting the focus from renewable energy systems – and thus new build plants and systems – to actual existing power plants and existing production facilities. Lead user identification was needed to be validated since the gap within the project and the trends influenced lead user classification.

On-site visits (3.2) were prepared by attending the Hannover Fair 2011 to reflect the actual state of the art in IIT security. This confirmed previous findings. Following on-site visits included a research power plant and an educational laboratory for energy transmission. During these visits, the project team learned about standardised processes, live network traffic, and solutions concepts from related research fields. This was extended by interviews with identified lead users.

Generation of *preliminary solution concepts* (3.3) prepared the lead user workshop and phase 4 of the traditional lead user method. The preliminary solution concept was written down in a draft of the *white paper* "ESCI – Enhanced Security for Critical Infrastructures" that was released on August 23rd 2011.

5.3.5. Phase 4 – Improvement

The *administration* of the workshop (4.1) was parallel work next to steps 3.2 and 3.3. The project team planned the workshop as a 2-day setting with a social event at the end of the 1st day. All identified lead users were invited, but had to proof own investments. Overall, five lead users and (now) 12 members of the project team (including the moderator) attended the workshop. The workshop was scheduled on August 25th and 26th 2011. The workshop was hosted by IHP GmbH.

The main task to *refine the concept* (4.2) was based on feedback to the previous mailed ESCI white paper 1.0 and incorporated lead user solutions into the product concept. The 1st day of the workshop introduced the overall project and clarified related questions. This day was intended to *foster a common understanding of the project*. This led to an open discussion about security constraints, the application field and its requirements, and the need for IT security in general. On the 2nd day, the workshop *group was separated into two* groups that covered challenges and solutions in (1) the Ethernet level and its interfaces and (2) the fieldbus level (see in reference to Herstatt/von Hippel 1992). Each group consisted of eight members.

Fundamental challenges to perform IT security-related operations were structured into detailed *sub-issues*. The results were discussed and merged after group work in the afternoon of the 2nd day. All participants discussed the implications from both groups and defined *possible configuration and application scenarios* to foster a later prototype. At the end of the workshop, the team had generated multiple product specifications to create a possible prototype. All specifications for the prototype were characterised by *real-time applications*.

A final *business case* (4.3) was generated by (1) an on-site visit at BTC AG – the developer of the PRINS process information system – in February 2012, (2) by reflecting the state of the art at the CEBIT fair in March 2012, (3) and a 2nd workshop in June 2012 with the focus on financing a spin-off. A (4) further meeting with the VISA project of DeCoit GmbH in August 2012 and (5) participation in the working group KRITIS of GI and CAST verified the specification. A 3rd workshop with the topic of CRITIS was scheduled in October 2012 and presented the idea to a crowd of experts from this field for validation reasons.

The distribution to the *senior management* (4.4) was immanent since the senior management was an active part of the lead user method.

The resulting concept is given in the 2nd version of the white paper and was presented at the Hannover Fair in April 2013 to verify compatibility with the markets of energy and industrial automation. The result was a conceptual model that was further implemented in prototype components to address multiple research fields and problem areas (in reference to a structured analysis).

Table 48 ESCI Components and their Descriptions

Component	Description of Components
Protocol analysis	Analysis of Profinet IO and implementation to identify specific IT security vulnerabilities (DoS, MITM).
Deep-packet-inspection	Implementation to decode and identify network messages (read-request, write-request etc.) Using the open-source IDS snort in Profinet IO networks; implementation of a support vector machine (serves as i_{02} in chapter 5.4).
Simulation	Simulation of a Profinet facility to test system components.
ESCI editor	Analysis and planning of an Industrial IT network using AutomationML. This enables further penetration tests (serves as i_{05} in chapter 5.4).
Java expert system	Implementation of an independent java-based rule engine for low-performance devices.
Distributed network sensors	Distributed architecture to identify security-related events. Implementation of an event scanner, event associator, and message dispatcher.
Communication	Authentication with cyaSSL for IDS components and shortECC for sensors (serves as i_{07} in chapter 5.4).

Table 48 provides an overview of the resulting ESCI components. The Java-based rule engine, a deep-packet-inspection component for Profinet, and the implementation of a lightweight ECC-algorithm (shortECC) for low-power devices like sensors and actuators outperform the state of the art and provide major advantages for the application field. The project officially ended in March 2013. Overall, challenges were encountered in the application of the traditional method.

5.4. Application of the Preference-Driven Lead User Method

5.4.1. Phase 1 – Preparation for Project Launch

The application of the Preference-Driven Lead User Method was done in the beginning of the year 2013 and relied partly (like in phase 1) on previous work with the traditional lead user method. The team foundation (1.1) was skipped since the team was already performing. This circumstance will apparently change in an exclusive application of the Preference-Driven Lead User Method. The project schedule (1.2) was confirmed and addressed the same business fields and target markets. The master project plan (1.3) was also adopted although the available time to apply the method became irrelevant. Further, this provides comparability.

5.4.2. Phase 2 – Identification of Trends and Internal Contributions

The 2nd phase incorporated previously identified trends (2.1). The trends from the traditional lead user method (see table 47) were re-evaluated for their relevance. For example, the *convergence of energy and industrial automation systems* was confirmed since the technical process continued over the past recent years. Also highly sensible trends like BYOD (bring your own device) and an increasing wireless connectivity in CRITIS were still relevant. However, the evaluation of trends was not necessary and was exclusively performed to confirm that the application of the Preference-Driven Lead User Method took place in the same environment.

The definition of an attribute set (2.2) to foster preference measurement was derived from market available solutions for IIT security. These analysed products from CISCO, Fortinet, and Siemens among other/additional Unified-Threat-Management (UTM) and Security-Information-and-Event-Management (SIEM) systems. The derived attributes were: (1) distributed logging, (2) deep packet inspection, (3) detection based on signatures, (4) detection based on anomalies, (5) authentication of network nodes, (6) encryption of network traffic, (7) a firewall application, (8) automated blocking, (9) individualised routing, (10) individualised network structure, and (11) radio field planning.

The identification (2.3) of internal contributions was also related to the traditional method. The contributions were: (1) using honeypots, (2) mechanisms to detect false and faked system events, (3) jamming detection, (4) distributed analysis, (5) identification and (trust) evaluation of new network components, (6) local analysis of critical events, (7) automated switching of authentication and encryption algorithms, (8) group-based key exchange, (9) voting mechanisms to react on incidents, and (10) mechanisms to handle BYOD.

5.4.3. Phase 3 – Exploration of the User Community

The 3rd phase of the applied Preference-Driven Lead User Method started with the classification of lead users (3.1). The construct of *lead userness*, items of the trend setting scale, and an own idea development were employed to foster this (see in relation to subchapters 2.2.3 and 2.3.3). The resulted classification consisted of six items to define lead users' high expected benefit and three items to define lead users being ahead of trends (see table 49).

Table 49 Employed Items to Reflect Users' Lead Userness

Dimensions of Lead Userness	Items Used to Generate Dimensions
High expected benefit	Market available security products do not fulfil my personal requirements to professional IT security.
	In my opinion, a detailed reflection of IT security reveals unsolved problems.
	The lack of IT security awareness in companies irritates me.
	I am confronted with problems of IT security often that cannot be solved with market available products.
	I am dissatisfied with some components of available IT security solutions.
	In the past, developer of IT security solutions were not able to solve my IT security-related problems.
Ahead of the trend	I am the one who discovers new trends, developments and challenges of IT security first in my social network.
	I have gained benefit from an early usage of new IT security solutions.
	I take part in IT security audits or hack challenges often.

This two-way classification was done because of a potential self-evaluation bias (as mentioned by Belz/Baumbach 2010) along with a general uncertainty within the project team to identify lead users without respect to an own idea development.

The survey design (3.2; CAWI) used the cover-page of the survey to foster the motivational introduction. This is given below (figure 16) and informs the respondent about the *survey intention*, the *technological background* of the problem, and the survey process. The cover page further reveals that the project was publicly funded and offered incentives like (1) a fictional thriller and (2) an invitation to the CRITIS workshop of the German Informatics Society (follow up communication with selected respondents revealed that the possibility to attend the workshop was a stronger incentive to take part in the survey than the book).

Figure 16 Cover Page of the IT Security Survey

The 1st contextual part of the survey covered the *lead user classification* using the lead userness construct and a 6-point rating scale (see figure 17).

The 2nd contextual part asked about a ranking of *state of the art attributes* (using 1-staged traditional self-explicated measurement) of IT security solutions and introduced the *contributions* of the project team with a binary rating. The 3rd part asked about an *own idea development* with detailed questions about the problem and later served as a verifier to select lead users. The 4th part performed the *holdout test* using three alternative stimuli: (1) a reactive solution, (2) a passive solution, (3) a market available solution for IIT security to verify the methodological validity. The questionnaire finishes with *demographic items* (5th part) and an option for feedback and *recommendation*. Figure 17 and figure 18 provide insights into the graphical design of the survey (in German language). Previous research in online social networks shows that reputation correlates with involvement (see Benjamin/Chen 2012). It was therefore to assume that detected lead users would recommend the survey within the IT security community.

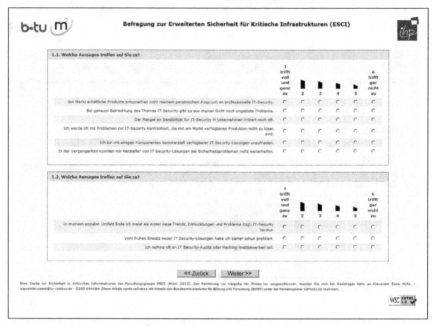

Figure 17 Classification via Lead Userness in the IT Security Survey

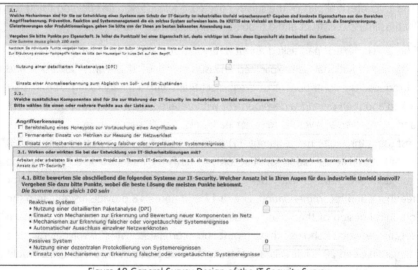

Figure 18 General Survey Design of the IT Security Survey

The survey distribution (3.3) was intended to cover a wide range of the market for IIT security. Thus, the distribution covered (1) *online distribution e.g. in social networks* etc. and (2) *offline acquisition*. The online distribution was done using the traditional screening approach within posting an invitation to the survey in social networks and online discussion boards. Table 50 summarises online activities.

Table 50 Online Activities to Distribute the Survey

Distribution Channels	Selected Examples	Amount of Contacts
Social networks	Xing, LinkedIn	54,834 users in 29 groups
Online discussion boards	Hackspaces, heise, gulli, SPS-forum	126,456 registered members in 39 boards
E-mailings	CAST e.V., IHK, lists.tu-darmstadt.de	Approx. 2,350 subscribers
Websites	www.esci-vrs, BTU Cottbus, IHP GmbH	Approx. 2,000 impressions
Internet relay chat	abjects, criten, efnet	Approx. 30,000 connections
Overall addressed contacts		Approx. 215,640 contacts

This was extended by *personal invitations* following a pyramiding approach (pyramiding shows benefits when used in poorly mapped search spaces as given by Poetz/Prügl 2010). This was also implemented in the traditional offline world. The project team called experts in this field and received referrals to whom an invitation was sent by e-mail after a first personal contact.

5.4.4. Phase 4 – Survey and Preference Measurement

The survey generated n=246 respondents. Among them were *n=111* respondents with sufficient data quality (4.1) measured by their external validity (minimum required Spearman correlation r=0.5) using a set of holdouts. This led to an *average Spearman r=0.77* that indicates sufficient validity for the measurement of state of the art attributes. The dimensions of the lead userness (see table 49) reflect "high expected benefit" (Cronbach's Alpha=0.80) and "being ahead of trend" (Cronbach's Alpha=0.72). The Pearson correlation between the lead userness (measured as the sum of all items) and the leading edge status (in reference to subchapter 2.3.3) indicates a sufficient fit (r=0.81, p<0.01). The correlation between an own idea development and the leading edge status (r=-0.367, p<0.01) as well as the correlation between an own idea development and lead userness (r=-0.308, p<0.01) indicate that there might be a *self-evaluation bias* in both directions (over evaluation and under evaluation).

Overall, n=43 respondents reported an *own idea development* (including conceptual thoughts). Among them were n=25 respondents with further developed ideas. In sum, n=11 (about 21% of all respondents with an own idea development) respondents wanted to reveal their ideas freely – one contribution was skipped. This is in line with previously introduced literature (see von Hippel/von Krogh 2006). The average job experience of all respondents was 11 years. They characterised themselves as (1) 68.4% being an IT administrator or manager in a business surrounding, (2) 36.9% being enthusiastic about the topic of IT security, and (3) 90% being related to the field of CRITIS. The respondent structure is (4) 90% male and 10% female.

The *analysis of the preferences* of the state of the art attributes (4.2) followed. Table 51 provides an overview of the importance per respondent group. The project team decided to identify lead users based on their own idea developments and internal screening for novelty. The results indicate significant differences in the evaluation of selected attributes (detection based on signatures, $p<0.05$; firewall, $p<0.05$; automated blocking, $p<0.1$).

Table 51 Overall Importance Values for SOTA Attributes

Attribute	Overall Importance Values (Std. Dev.)		
	Overall (n=111)	Lead user (n=14)	Non-lead user (n=97)
Encryption	14.27 (8.15)	16.49 (10.97)	13.95 (7.68)
Individualised network structure	7.97 (7.95)	11.04 (12.12)	7.53 (7.14)
Distributed logging	8.30 (6.86)	9.57 (7.98)	8.11 (6.71)
Deep packet inspection	6.48 (5.77)	4.72 (5.14)	6.74 (5.83)
Radio field planning	3.23 (4.34)	4.02 (5.32)	3.11 (4.20)
Automated blocking	6.41 (6.97)	7.49 (5.95)*	6.25 (7.12)
Detection based on signatures	9.74 (8.38)	4.82 (4.53)	10.45 (8.58)**
Individualised routing	6.48 (6.77)	4.76 (5.21)	6.73 (6.96)
Detection based on anomalies	9.69 (7.48)	9.06 (4.38)	9.78 (7.84)
Authentication	13.07 (8.62)	16.14 (12.03)	12.63 (7.99)
Firewall	14.35 (8.71)	11.90 (6.97)	14.71 (8.90)**
Significance: *$p<0.1$, **$p<0.05$, ***$p<0.01$			

The analysis of internal contributions (4.3) showed partial acceptance by the user community. Interestingly, no significant differences in the evaluation structure of lead users and non-lead users could be measured. Table 52 illustrates the results.

Table 52 Acceptances for Internal Contributions

Internal Contribution	Overall Importance Values (Std. Dev.)		
	Overall (n=111)	Lead user (n=14)	Non-lead user (n=97)
Honeypots	0.56 (0.50)	0.63 (0.49)	0.55 (0.50)
Detection of faked/ false system events	0.60 (0.49)	0.57 (0.51)	0.61 (0.49)
Jamming detection	0.51 (0.51)	0.34 (0.48)	0.54 (0.50)
Distributed analysis	0.38 (0.49)	0.33 (0.47)	0.39 (0.49)
Accreditation of new components	0.60 (0.49)	0.71 (0.47)	0.59 (0.50)
Local analysis	0.21 (0.41)	0.10 (0.28)	0.23 (0.42)
Adaption of algorithms	0.68 (0.47)	0.64 (0.50)	0.68 (0.47)
Group-based key exchange	0.48 (0.50)	0.34 (0.48)	0.49 (0.50)
Voting system	0.15 (0.35)	0.23 (0.42)	0.13 (0.34)
BYOD	0.44 (0.50)	0.62 (0.49)	0.41 (0.50)
Significance: $*p<0.1$, $**p<0.05$, $***p<0.01$			

Previous chapters point to the preferability of *similar evaluation structures* for both user groups. Franke/von Hippel (2003) showed that respondents had heterogeneous desires, but provided homogeneous judgement of the importance of the security-related functions.

Collaborative filtering (4.4) provided information of the evaluation of external contributions using all respondents. External Contributions were (e_{01}) IPS/IDS algorithms for field bus devices, (e_{02}) a central analysis of system events by multiple network nodes in a distributed network, (e_{03}) a smartcard based VPN system, (e_{04}) a permanent usage of metrics to monitor network load, (e_{05}) a secure code attestation with specific requirements, (e_{06}) a redundant architecture of information systems, (e_{07}) rate limitation, (e_{08}) usage of a graphical firewall, (e_{09}) 'secure communication per default', and (e_{10}) E-DRM. No significant differences were observed between both user groups.

5.4.5. Phase 5 – Lead User Workshop

The project team continued with the pre-assessment of the contributions (5.1). A brief patent review revealed that multiple IP issues are from concern. The project team scanned, evaluated, and categorised the patent into minor relevant, relevant, and major relevant. Minor relevant patents can be addressed by (simple) workarounds. Relevant patents need to be addressed by more sophisticated effort and partnerships. Major relevant patents cannot be addressed.

The preparation of the workshop (5.2) and the concept generation (5.3) followed in *reference to the traditional method*. The resulting product concept for a next generation IIT security solution is named System Pref and is determined by code attestation (e_{05}), a permanent use of metrics to measure network load (e_{04}), dynamic adaption of authentication and encryption algorithms in case of cyber-attack (i_{07}), and detection of false and fake system events (i_{02}).

A re-test (5.4) to compare the derived system with the result from the traditional lead user workshop (System ESCI; see subchapter 5.3.5) and with a market available solution (System Mark) shows that 'System Pref' is in favour and has significantly ($p<0.1$) higher market potential. In contrast, the novelty is reduced (see table 53). The re-test employed constant-sum measurement for the rating. It further used a 10-point rating scale to determine market potential and novelty. An analysis between both user groups revealed no significant differences.

Table 53 Performance Measurement of Innovative IT Security Solutions

System	Mean (Std. Dev.)	Market Potential (Std. Dev.)	Novelty (Std. Dev.)
Mark	30.32 (14.88)	5.55 (2.16)	5.10 (1.83)
ESCI	33.87 (8.44)	5.65 (2.04)	5.06 (2.03)*
Pref	35.81 (12.66)	6.23 (2.06)*	4.68 (2.17)
n=33; Significance: *p<0.1, **p<0.05, ***p<0.01			

In reference to Urban/von Hippel (1988), this similar evaluation structure confirms the expected market potential for the derived IT security solution. The system is then forwarded to the senior management (5.5).

5.5. Implications

This chapter briefly introduces the application field of IIT security which serves for the empirical analysis of the Preference-Driven Lead User Method. The field of IIT security is in need of innovation since traditional IT security mechanisms tend to increasingly fail, as the examples of Stuxnet, Flame, and Duqu showed. These traditional mechanisms are for example firewalls with white- or blacklisting of communication partners. Modern IT security mechanisms for authentication, encryption and Intrusion Prevention are not standard practice since todays IIT systems deal with low performance and high energy consumption. Thus, computational operations are costly. This becomes extremely challenging in the field of Critical Infrastructures where IT security related incidents are constantly rising. The project "Enhanced Security for Critical Infrastructures" (ESCI) addressed this topic by proposing a distributed peer-to-peer approach. The future Intrusion Prevention System ought to use available computational resources within a network to perform distributed analysis of network traffic.

The project team applied the traditional lead user method using pyramiding. The Preference-Driven Lead User Method was applied in comparison and made use of similar respondents (lead users) and additionally non-lead users. In sum, both methods were able to provide promising results. The re-test compared both approaches in terms of market potential, novelty, and relevance. The derived concept from the Preference-Driven Lead User Method showed significantly better results in terms of market potential. The derived concept from the traditional lead user method was rated significantly better in novelty. This was expected.

However, both methods were applied in an environment with limited resources and had to respect implications of a publicly funded research project. This was for example a time gap within the traditional method. It could further be observed that the identification of lead users using the lead userness construct was not necessarily highly correlated with the own idea development. Further, the demand for data quality within the survey required to drop respondents. Although the respondents provided own ideas to the survey, their preferences had to be skipped in further analysis in reason of inconsistency.

6. Empirical Verification of Strengths and Weaknesses

6.1. Overview

Academic applications of the lead user method point to several challenges in practice, which could be derived from theoretical input and observations in chapter 2. The present chapter deepens the understanding of *practical strengths and weaknesses and gives insights into the automotive industry, the mechanical engineering industry, and the market intelligence industry in Germany.* The empirical study addressed n=874 corporate members with n=311 usable questionnaires and led to n=84 respondents, who were familiar with the lead user method. This points to a familiarity with the method of 27% and is in line with previous findings in literature. Overall, 18 basic statements were derived from theory to reflect the respondent's attitude towards the usage of the lead user method in innovation projects. Besides, each industry study was accompanied by expert interviews to reframe preliminary statements and to verify their applicability.

The overall analysis concentrates on *descriptive results, namely the comparison of mean values, investigates the correlations between the statements, and OLS regression* analysis to highlight major influences on the decision to apply the lead user method if applicable. In general, the results show evidence of methodological weaknesses that influence the decision to use the lead user method. On the one hand, the method is able to provide benefits as suggested by theory, but faces restrictions inside the firm like NIH, time and cost constraints and missing trust, on the other hand. The results vary across the chosen industries, which can be influenced by the general character of the industry itself, the structure of SMEs and their influence to the market as well as specialised products. In sum, it is shown that *actual users and non-users evaluate the lead user method differently.*

The remainder of this chapter introduces the methodological background along with the basic statements that were derived from literature, and presents the three industry samples. The presentation deepens the understanding per industry in a separate subchapter and ends with cross-industrial findings. Overall, theoretical assumptions that led to the concept of the Preference-Driven Lead User Method were confirmed by practical insights. This chapter ends with a preliminary evaluation of the new method in these industrial surroundings.

6.2.   Verification in the German Automotive Industry

6.2.1. The Application Field of Automotive Development

The 1st case study is done in the automotive industry. This industry is considered to be the *main driver of innovations* and is one of the leading industries to foster Germany's economic wealth (see Destatis 2013 for underlying information on export revenue, FDI and number of employees). It covers more than 1,200 enterprises with about 1 Million employees and generated a turnover of about EUR 413 Billion in 2012 (see Destatis 2014 in reference to the national classification of economic activities groups WZ08-29 and WZ08-30; the whole industry cannot be completely mapped, because of several specialised enterprises that work for multiple industries but are mainly associated to other groups).

The need for innovation in this industry became more present in the *economic crisis* of 2008 and 2009, which hit the automobile industry hard and led to a massive decrease of the sales volume in the US and European market. Overall, the German Automotive Original Equipment Manufacturers (OEM) invested EUR 34.8 Billion in research and development in 2011, which represents 9.4% of the industry's turnover and is an annual increase of approximately 16% (ZEW 2013).

Further, the *length of product development life cycle* is decreasing while development costs are increasing (von Corswant/Fredriksson 2002). Today, this industry is mainly influenced by globalisation that improves market prospects and cost advantages, but also *competition* (Archibugi/Iammarino 1999). The automobile model life cycle has shortened drastically from about 10 years in the 1980s to about five to six years after 2000 (Maxton/Wormald 2004 and Dannenberg 2005). For example, the Japanese automaker Nissan has unveiled its strategy 'Nissan Power 88' that aims to launch an all-new model every six weeks (Ghosn 2011). A strategy like this can be observed in many automotive enterprises, which leads to a massive market penetration and requires a fundamental change in business models, brand politics and manufacturing processes (see for example Volkswagen 2010 for investments in home power plants and Volkswagen 2009 for an early description of its MQB platform to aggregate production).

Those developments are linked with maximum *effort on the suppliers' side* in terms of costs management and supply-chain management (see for example Meißner 2012 for a brief description of implications for automotive suppliers and

Richter/Hartig 2007 for a possible network approach that was done by BMW Group). Moreover, the greater competition between the international automobile companies (Balcet et al. 2012) to achieve customers' satisfaction highlights the need for innovation in multiple aspects, like in quality and additional functions.

Literature shows that the *traditional structure of their customers* changed drastically (see Feldmann/Armstrong 1975 for an early identification of early adopters in the automotive industry) and manufacturers need to address a diverse market with multiple preference structures.

However, innovative demands are generated by multiple *stakeholders* within this industry, e.g. by customers and by national law. The latter one is for example influenced by climate regulations (see e.g. Eggers/Eggers 2011 for climate trends in this industry and Freyssenet 2011 for operational scenarios to achieve them and their implications for the industry). Those demands are either shifted from the automotive manufacturer to suppliers and engineering service agencies or needed to be addressed by a solution that is developed in-house (see Fixson et al. 2005 for a case study in modularisation and outsourcing, and Rese et al. 2015 for an update of innovation structures in the industry).

Nevertheless, innovations often fail to be *financially profitable*. One reason can be seen in unreliable approaches for sales forecasting (see for example Urban et al. 1990 for a discussion with 46% difference between the actual and the predicted sales volume). Thus, innovation costs cannot be refunded by mass production. Further, Pauwels et al. (2004) showed that major innovations in this industry face market uncertainties. The authors noticed that a major innovation can also have a serious negative impact on the enterprise's long-term financial income and firm value. This can be the case if an overall new platform is traduced that will not fit customers' preferences, their needs and cannot satisfy the forecasted data (see e.g. Hillenbrand 2007 for the historical Edsel case). In contrast, the market offers much potential. Still, forecasts expect that the "... total number of household vehicles in 2025 will reach 235 million, representing a 31% increase over the 25 years" (Feng et al. 2011, p. 593) for the United States.

Thus, *user and supplier innovation* faces high potential in this industry (see for example Ili et al. 2010 for a brief introduction to the drivers of open innovation in the automotive industry). The opening of the innovation process for the automotive industry is "... the enlargement of company's competence base, the stimulation of creativity and capability of generating new ideas, the reduction and sharing of risks related to innovation activities and costs of innovation process" (Lazzarotti et al. 2013, p. 53). Academic studies have already highlighted the usage of lead users in this industry and found that the lead user method is not the 1st methodological choice to generate innovations in general (see for example Lichtenthaler 2004 for a study with seven European and North American enterprises). Lead users are frequently used to generate radical innovations that are mainly driven by regulatory environments.

One might argue that those innovations will be no unique selling proposition for the enterprise in the future and thus *lead users provide a convenient way to adapt solutions from analogous markets*. Lichtenthaler (2005) showed for the same sample that lead users were also used to manage market uncertainties. Notably, Lichtenthaler (2005) mentions that in the late 1990s "... the interest in managing uncertainty concerning the outcome of R&D projects and of external trends in general increased..." was a trend (Lichtenthaler 2005, p. 392). This was addressed by using the all-new lead user method by practitioners. It is in contrast with the theoretical findings in literature on weaknesses, especially to address market uncertainties and provides a hint that this approach might have failed in the past.

Practical case studies like a lead user project at Webasto (see e.g. Lang 2006 and Lang/Reich 2008) are known in this field. Hering et al. (2011) showed a link to the necessary opening of the innovation process that is in contrast to high demands on secrecy. A further story of the lead user method in the field of *future mobility* is provided by the German project BeMobility. BeMobility is concerned about electro-mobility in the greater Berlin area and was coordinated by Deutsche Bahn from 2009 to 2011. The lead user method was applied with the focus on Berlin solely. Thus, the *local search bias* may be present. Trend analysis was done by the identification of specific local areas in Berlin, which are characteristic for specific social changes. Respondents who were living or working in specific areas were interviewed and observed. The results led to the modelling of a specific reference respondent for each area – called Persona – and framed criteria to perform lead user identification. Important trends were identified as co-working spaces, as socially connected and active elderly people, and as integrative multi-generation living concepts. The integration of lead users was separated by the identified trends and was done in three workshops (one workshop for each trend). This unusual process led to ideas and concepts (cf. Wolter/Knie 2011).

6.2.2. Survey Preparation and Data Collection

The survey was accomplished during the summer and autumn of 2012 and started with seven *preparing expert interviews* in July conducted with undergraduate assistance. The expert interviews were helpful to become more familiar with the market and to gather first hand experiences to frame the later questionnaire around the core statements from table 20.

Overall, the results of the expert interviews that covered multiple areas within the value chain of the German automotive industry, showed that the lead user method is majorly unknown in this industry and is seldom used. The interviews pointed to several aspects that argue about the benefit of the lead user method. One aspect is described by the *general unwillingness to open the innovation process* and the *awareness of confidential product developments*. This is in line with basic literature (see e.g. Chesbrough/Brunswicker 2014). Further, the integration of customers in this field is mainly understood as performing preference measurement and to test the acceptance of concepts that have been developed. This has historical reasons and is accompanied by an internal innovation development and a mature NIH (see Katz/Allen 1982 for a detailed introduction to NIH and its sustained affects).

Worse, experts agreed on the fact that *engineers* from this field assess themselves to be able to *provide solutions for customers' needs and market trends*. This mindset leads to a non-usage of the method and a sustained lack of trust in the method. Another aspect is described by the intention of the method to develop radical innovations. History shows that the *automotive industry is mainly driven by incremental innovations*. Radical changes, like the introduction of non-combustion engines to baseline models, are somewhat new to this market. The willingness to develop incremental innovations is necessary to achieve *scale effects* and enable an accelerated refund of the investments due to mass production.

Further aspects emphasise that innovations are triggered by functional problems with already existing components (e.g. a component got broken) and are based on already surveyed reliable customers' preferences. The argumentation of high expenses to perform the lead user method, the binding of resources, also hinders the usage of the method in accordance to the expert interviews.

Table 54 Used Statements in the Automotive Sample

A1: Lead users possess competencies that would enable a development of new concepts.
A2: Lead users have no strong interest that their own needs will be fulfilled.
A3: Lead users are able to estimate product requirements in advance to ordinary customers.
A4: Lead user needs will reveal future needs of ordinary customers.
A5: Lead users provide ideas that are unknown yet.
A6: Lead user contributions are accepted by our engineers and developers.
A7: The Lead User Method is less costly in comparison to other methods of IM.
A8: The Lead User Method is less time consuming in comparison to other methods of IM.
A9: The trust to integrate lead users as the centre of innovation is given.

Summarised, the lead user method provides strengths and faces weaknesses in the automotive industry. Table 54 reflects the statements that were derived from the expert interviews used within this survey. The survey itself started in autumn 2012. The distribution of the survey was done with the help of several automotive associations and other promoters like the VDA, IHK, Dekra, ADAC, ACE, TÜV Nord, VWI and others. The present analysis also includes laggards.

6.2.3. Descriptive Results and Regression Analysis

The survey generated n=112 usable questionnaires out of 295 respondents. This leads to a response rate of about 38.0% and to a general response rate of about 17.4% in calculation with an estimated amount of 1,696 OEM and further downstream enterprises in the German automotive industry (in reference to Vollrath 2002 and Destatis 2014). The respondents can be categorised in 9.8% OEMs, 54.5% suppliers from any tier, 5.4% consulting enterprises, and 19.6% engineering service providers. The other 10.7% represented logistic services, tuner, commercial vehicles, and motor bikes as well as other fields. Overall, the *sample consists of 75.0% SMEs*. That is in line with Schade et al. (2012) who found a similar distribution of the company sizes in their analysis of automotive suppliers. Table 55 summarises the respondents and their structure.

The term "lead user" was familiar to 32 respondents (28.6%). Further, the lead user method was or is used in 23.2% of all cases (26 out of 115 cases). When the lead user method was familiar, then it was often used (81.3%). *This is in contrast to the expert interviews*. The main role of the method was seen to generate incremental innovations in about 62.5% of all cases. The focus is on product innovation with 71.4%.

Table 55 Structure of the Respondents from the Automotive Sample

Descriptive Aspects	Sample Share
Amount of usable respondents in relation to the market	112 (6.6%)
Share of respondents being familiar with lead users	32 (28.6%)
Usage of the lead user method	26 (81.3% of experts)
Description of the Methodological Application (Multiple Answers Possible)	
Search for radical innovation	37.5%
Search for incremental innovation	62.5%
Appliance in process innovation	25.0%
Appliance in product innovation	71.4%
Special appliance for fundamental concepts	31.3%
Special appliance in automobile equipment	21.9%
Special appliance for product line optimisation	21.9%
Description of the Respondents' Enterprise (* as per EU Definition)	
Small-sized enterprises	20.5%
Medium-sized enterprises	54.5%
Greater enterprises	25.0%

The Chi-Square-test showed significant differences for statement "Lead users are able to estimate product requirements in advance to ordinary customers." (a3) and statement "The Lead User Method is less costly in comparison to other methods of IM." (a7) ($p < 0.1$) under assumption of normal distribution. In fact, the trust to incorporate lead users (a9) was found to possess significant ($p < 0.01$) differences between both user groups.

Table 56 indicates that the lead user method has potential in this application field and receives positive evaluation, but faces weaknesses related to the statements a2, a7, a8, and a9. This is in line with findings of Rese et al. (2015) and Creusen et al. (2013). They found that the evaluation of costs, time, and confidence is in significant negative correlation with the usage frequency. In contrast, users and non-users evaluate lead user contributions and the potential to predict the needs of ordinary customers to *cover a broad market* positively. This is in line with the theory (see von Hippel 1986) and partially fits the results of the expert interviews, although lead users have a strong interest to solve their own needs (a2). However, statement a9 points to the trust to integrate lead users as the centre of innovation and shows strong significant ($p < 0.01$) differences between non-users and users.

This leads to the assumption that if the lead user method was used once, then the *benefit outperforms the unwillingness to open the innovation process* at all. This is also partially true for statement a7, but shows weak significant ($p < 0.1$) differences. In contrast, statement a3 claims that lead users are able to estimate product requirements in advance to ordinary customers, but this shows a different evaluation between non-users and users (p<0.1), who rate this statement negative.

Table 56 Results per Respondent Group in the Automotive Sample

Strengths and Weaknesses of the Lead User Method in Practice	Respondents Being Familiar With the Lead User Method	
	Not applied n=6 Mean (Std. Dev.)	Applied n=32 Mean (Std. Dev.)
A1: Lead users possess competencies that would enable a development of new solution concepts.	3.17 (0.98)	3.73 (0.87)
A2: Lead users have no strong interest that their own needs will be fulfilled.	2.00 (1.10)	1.85 (0.83)
A3: Lead users are able to estimate product requirements in advance to ordinary customers.	4.17 (1.17)*	4.04 (0.77)
A4: Lead user needs will reveal future needs of ordinary customers.	3.17 (1.17)	3.12 (0.77)
A5: Lead users provide ideas that are unknown yet.	3.50 (1.05)	3.67 (0.98)
A6: Lead user contributions are accepted by our engineers and developers.	3.17 (0.98)	3.23 (0.99)
A7: The Lead User Method is less costly in comparison to other methods of Innovation Management.	2.50 (0.84)	3.12 (0.91)*
A8: The Lead User Method is less time consuming in comparison to other methods of Innovation Management.	2.17 (0.98)	2.88 (0.91)
A9: The trust to integrate lead users as the centre of innovation is given.	2.33 (1.37)	3.35 (0.80)**
Scale of mean value: 1 (totally disagree) ... 5 (totally agree); Significance: *p<0.1; **p<0.01		

A detailed view on the Pearson correlations between the statements reveals significant relations within this sample. Table 57 provides the correlation matrix. The correlation between statement a1 and statement a5 (r=0.35, p<0.1) proves the *general benefit that the lead user method is assumed to have*. The correlation between statement a2 and statement a3 (r=-0.61, p<0.01) confirms the basic theory on lead users. Moreover, statement 5 fosters this argumentation and is in significant correlation with statement a5 (r=0.38, p<0.05), statement a6 (r=0.36, p<0.05), statement a7 (r=0.42, p<0.05), and statement a8 (r=0.33, p<0.1).

The benefits of the method seem to justify the efforts in this industry. Further, the main aspect is defined by the common trust to use the method.

This trust depends on several *prejudices*. Lead user contributions are accepted by internal engineers and developers and are necessary to develop an *enterprise-wide trust* to use this method, as expressed by the correlation between a6 and a9 (r=0.53, p<0.01). The evaluation of the cost expenditures is in line with the estimated time that is needed to apply the method ($r=0.41$, $p<0.05$). If trust can be raised, then the willingness to accept higher expenses is strong ($r=0.39$ for costs and $r=0.35$ for time, $p<0.05$).

Table 57 Inter-Correlation Matrix of Statements in the Automotive Industry

	(A1)	(A2)	(A3)	(A4)	(A5)	(A6)	(A7)	(A8)
(A2)	-0.23	1.00						
(A3)	0.29	-0.61 ***	1.00					
(A4)	0.24	0.20	0.08	1.00				
(A5)	0.35 *	0.02	0.19	0.38 **	1.00			
(A6)	0.24	0.11	0.10	0.36 **	0.13	1.00		
(A7)	0.27	-0.00	0.04	0.42 **	0.04	0.29	1.00	
(A8)	0.26	-0.00	0.14	0.33 *	0.14	0.20	0.41 **	1.00
(A9)	0.24	-0.09	0.03	0.29	0.16	0.53 ***	0.39 **	0.35 **
Significance: *$p<0.1$, **$p<0.05$, ***$p<0.01$								

The ordinary least squares regression (OLS) is used to analyse the internal structure that leads to the usage of the lead user method within the automotive industry. The usage was rated on a 5-point Likert scale with 1 (usage in no project) to 5 (usage in every project). Table 58 shows the results of the OLS regression which revealed significant influences by statement a9.

This describes a lack of trust to integrate lead users per se as a major weakness within the automotive industry in Germany. This result is sufficient and the variance inflation factor shows acceptable values for all presented statements (mean 1.67). A further analysis using stepwise OLS confirmed the significant impact of trust (statement a9; $b=0.72$, $t=3.53$, $p<0.001$; $r^2=0.29$; $F=12.44$) to explain the usage of the lead user method, solely.

Overall, the automotive industry faces NIH that is highly responsible for a lack of trust in the method. *Trust is the most important aspect to incorporate lead users in the innovation process and can be built by actual usage only.* The benefits of lead users are known and are of importance, but the strong interest of lead users to pursue their needs exclusively (statement a2) hinders the methodological application along with time and cost constraints.

Table 58 OLS Regression Analysis for the Usage in the Automotive Industry

Strengths and Weaknesses of the Lead User Method in Practice (n=32)	OLS Regression Coefficient (Std. Err.)	T-Value
A1: Lead users possess competencies that would enable a development of new solution concepts.	0.23 (0.21)	1.12
A2: Lead users have no strong interest that their own needs will be fulfilled.	0.35 (0.26)	1.33
A3: Lead users are able to estimate product requirements in advance to ordinary customers.	0.11 (0.26)	0.43
A4: Lead user needs will reveal future needs of ordinary customers.	-0.09 (0.25)	-0.35
A5: Lead users provide ideas that are unknown yet.	-0.28 (0.19)	-1.43
A6: Lead user contributions are accepted by our engineers and developers.	-0.14 (0.21)	-0.65
A7: The Lead User Method is less costly in comparison to other methods of Innovation Management.	0.13 (0.22)	0.63
A8: The Lead User Method is less time consuming in comparison to other methods of Innovation Management.	0.28 (0.20)	1.43
A9: The trust to integrate lead users as the centre of innovation is given.	0.54 (0.22)	2.51**
R-Square: 0.50**; F-Value: 2.39; Significance: **$p<0.05$		

6.3. Verification in the German Mechanical Engineering Industry

6.3.1. The Application Field of Mechanical Engineering

The 2nd case study was done in the field of mechanical engineering. This industry is one of the most important industries in Germany, but is different in terms of customer-triggered developments (see Destatis 2013 and VDMA 2013). Thus, the mechanical engineering industry is not necessarily fostered on mass production, but on *customer-specific orders* and is described by the national classification of economic activities within group WZ08-28. The industry generated a turnover of EUR 221 Billion in 2012 and EUR 206 Billion in 2013 (see Destatis 2013 and VDMA 2014 for further details). The industry covers about 5,200 enterprises with about 1 Million employees (in reference to Destatis 2014, but this is vague in reason of the basic requirements to determine an enterprise by WZ08). Complementary studies report between 6,393 and 16,412 enterprises when small-sized enterprises with less than 50 employees are taken into consideration, too (see e.g. VDMA 2014 for the 1st and Destatis 2013 for the 2nd count). The overall investment in R&D activities was estimated to about EUR 5 Billion in the year 2011 (see Destatis 2013).

The *economic crisis* hit the mechanical engineering industry, too. The capacity utilisation plunged during the crisis to 67.5% in 2009 and regenerated to about 85% in 2012 and 2013 (median 86.2% during 2004 to 2012 in reference to VDMA 2013, p. 13 and VDMA 2014). The number of employees, which fell to 908,000 in 2010 and reached its peak with 995,000 employees in 2013, shows a similar perspective. Still, this industry faces decreasing income orders by -3.3% in 2012 and -2.0% in 2013. The German machinery export meanwhile increased by 5.1% in 2012 in comparison to the previous year and generated a turnover of about EUR 150 Billion with its key markets in Europe and Asia. This turnover was also achieved in 2013 (see VDMA 2014), although the important export to China decreased by -9.6% in 2012 and further by -3.3% in 2013. Today, the Chinese mechanical engineering industry is a major competitor next to USA and Japan in trade activities.

"In particular in Germany, the mechanical engineering companies had to cope with massive decreases in orders received in the summer, but they were able to catch up on that by the end of the year" (VDMA 2013, p. 16). Overall, the VDMA emphasised in accordance to the "ifo business climate survey" that the production in every fourth mechanical engineering company was negatively affected by decreasing orders (see VDMA 2014).

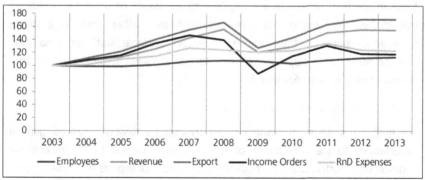

Figure 19 Economic Development in Mechanical Engineering (2003 to 2013)

(Reference: Illustration in Reference to VDMA 2007, VDMA 2010, VDMA 2014,

Wissenschaftsstatistik 2007, Wissenschaftsstatistik 2009, Wissenschaftsstatistik 2013)

Figure 19 shows that the global machinery market encounters an underlying regeneration with an increasing global competition. Herkommer (2013) reported a plunge of 18% in sales in 2009 in comparison to 2008.

Today, mechanical engineering faces multiple future trends. For example, Industry 4.0 (see e.g. Blanchet et al. 2014 for a practical insight) highlights the amplified *automation of engineering systems* and their *ICT components*. This trend is described along with the long-term demand for the *Internet of Things* (IoT) in which the phenomenon of user innovation, e.g. by toolkits, could already be observed (see for example Hribernik et al. 2011 and Cvijikj/Michahelles 2011 for early experiences). This trend affects internal systems of enterprises, their production processes and technical features of their products. Further, industrial automation is becoming a serious topic in many application fields of the industry's customers (Meinberg 2006 presented several application examples). This is accompanied by an increasing awareness of IT security in these Critical Infrastructures (see e.g. Stecklina/Sänn 2012). Other trends are described by the green revolution, new requirements for more efficient products, and *environmental friendly manufacturing processes*.

This requires innovations and customers' input. Previous studies pointed to a moderate familiarity (32.7%) with lead users and a *seldom use* (Mean value 1.39 out of 5 in reference to Altmann 2003) in this industry. Reger/Schultz (2009) showed that customer feedback is valuable for further innovations, but especially *SMEs in the field of mechanical engineering are aware of costs for innovation management* and prefer less expensive methods. Further, there is a strong dependency between the company size and available staff for innovation management. Thus, approaches need to be efficient despite the fact that *manufacturing is customer-specific*.

6.3.2. Survey Preparation and Data Collection

The empirical study started with preparatory expert interviews to reveal general expectations to the acceptance of the lead user method in the mechanical engineering industry. The expert interviews covered four experts from the regional Chamber of Commerce and Industry (IHK) and from the VDMA and was accomplished with assistance of undergraduates.

The interviews have shown that the mechanical engineering industry is mainly driven by *incremental innovations*, which is similar to the automotive industry. The traditional role of the customer in this industry can be described as revealing product-related problems and expressing their own (dis-) satisfaction with the solution. Customers are the main triggers of innovations in this field, but are not supposed to act as a partner in the field of innovation. Thus, user innovations by customers and by suppliers are seldom because of less specific knowledge. Moreover, product innovations are often combined with service enhancements.

The quantitative majority of this industry is represented by SMEs with less than 50 employees, which have specific requirements to methodological innovation management in general. Notably, the experts reported multilateral industrial partnerships among the enterprises and a basic trust within this industry - unlike in the automotive industry. Nevertheless, *product piracy is from concern* and occurs in forms of 1:1 product copies or reversed engineered processes. In addition, telephone interviews showed that there is no concern for patent issues, but product copies drive the topic of product piracy.

The development of the questionnaire was derived from the application field of the automotive industry. The questions were adapted to the new context of mechanical engineering. The study itself was hosted during autumn 2013 and was distributed with the help of undergraduate assistance and the VDMA. The basic statements on the contributions of lead users were adopted from table 20.

6.3.3. Descriptive Results and Regression Analysis

The survey generated n=99 usable questionnaires out of 204 respondents (48.5%). This leads to a general response rate of about 3.9% based on the actual market size (in reference to Destatis 2014). The sample consists of 29.3% small enterprises with less than 50 employees, 42.4% smaller medium-sized enterprises with less than 249 employees, 16.2% greater medium-sized enterprises with less than 1,000 employees, and 12.1% big enterprises (see Destatis 2013).

The term "lead user" was familiar to 19 respondents (19.2%). 16 respondents confirmed the usage of the lead user method (84.2%). Only 18.8% of the users applied the method to develop radical innovations. This is in line with previous literature and with results of the expert interviews.

The Shapiro-Wilks-test on normal distribution shows significant ($p<0.05$) results. The Mann-Whitney-U-test resulted in no significant differences. An additional Chi-Square-test under the assumption of normal distribution reveals significant ($p<0.1$)

differences in the evaluation of the costs for incorporating lead users (statement m7). Table 60 provides the mean values of the rating per user group.

Table 59 Respondent Structure in the Mechanical Engineering Sample

Descriptive Aspects	Sample Share
Amount of usable respondents	99 (2%)
Share of respondents being familiar with lead user	19 (19.2%)
Usage of the lead user method	16 (84.2% of experts)
Description of the Methodological Application (Multiple Answers Possible)	
Search for radical innovation	18.8%
Search for incremental innovation	81.2%
Appliance in process innovation	18.8%
Appliance in product innovation	81.2%
Appliance in line with technology push	50.0%
Special appliance for fundamental concepts	31.3%
Special appliance for product configuration	43.8%
Special appliance for technological innovation	37.5%
Special appliance for product line optimisation	50.0%
Description of the Respondents' Enterprise (* as per EU Definition)	
Small-sized enterprises	29.3%
Medium-sized enterprises	42.4%
Greater enterprises	28.3%

Table 60 Results per Respondent Group in Mechanical Engineering

Strengths and Weaknesses of the Lead User Method in Practice	Respondents Being Familiar With the Lead User Method	
	Not applied n=3 Mean (Std. Dev.)	Applied n=16 Mean (Std. Dev.)
M1: Lead users possess competencies that would enable a development of new solution concepts.	4.33 (0.57)	4.05 (0.62)
M2: Lead users have no strong interest that their own needs will be fulfilled.	1.67 (0.57)	1.67 (1.05)
M3: Lead users are able to estimate product requirements in advance to ordinary customers.	4.67 (0.58)	4.33 (0.62)
M4: Lead user needs will reveal future needs of ordinary customers.	3.67 (0.58)	3.56 (0.89)
M5: Lead users provide ideas that are unknown yet.	4.00 (1.00)	3.75 (0.68)
M6: Lead user contributions are accepted by our engineers and developers.	4.00 (1.73)	4.25 (0.97)
M7: The Lead User Method is less costly in comparison to other methods of Innovation Management.	3.33 (1.53)	3.67 (0.98)*
M8: The Lead User Method is less time consuming in comparison to other methods of Innovation Management.	2.67 (0.58)	3.56 (1.03)
M9: The trust to integrate lead users as the centre of innovation is given.	2.33 (0.58)	1.81 (0.54)
Scale of mean value: 1 (totally disagree) ... 5 (totally agree); Significance: *p<=0.1		

Overall, the lead user method and its contributions to the innovation management process are *positively evaluated* (see statements m1, m3, m4, m5, and m6). In contrast, users and non-users evaluated the *trust* (statement m9) and the *time consuming process* of the lead user method (statement m8) negatively. This is partially in line with findings in the automotive industry and identifies this as one of the important issues. In addition, statement m2 indicate that lead users are pursuing their own needs to be solved.

The analysis of the correlations of the statements showed that there is a positive significant positive correlation (r=0.47, p<0.05) between lead users' competencies and their ability to predict ordinary customers' needs. A further less significant positive correlation (r=0.43, p<0.1) could be seen between time and costs. Lead users are able to reveal unknown ideas but face a lack of trust to be incorporated in the innovation process (r=-0.47, p<0.05) in general.

Table 61 Inter-Correlation Matrix in the Mechanical Engineering Sample

	(M1)	(M2)	(M3)	(M4)	(M5)	(M6)	(M7)	(M8)
(M2)	0.07	1.00						
(M3)	-0.16	-0.34	1.00					
(M4)	0.47**	0.24	-0.19	1.00				
(M5)	0.15	-0.19	0.14	0.31	1.00			
(M6)	-0.36	-0.37	-0.02	0.14	-0.39	1.00		
(M7)	0.04	-0.25	0.13	0.39	0.07	0.00	1.00	
(M8)	-0.39	0.21	-0.05	-0.04	-0.02	0.04	0.43*	1.00
(M9)	-0.14	0.14	0.07	-0.22	-0.47**	0.12	-0.27	-0.21
Significance: *p<0.1, **p<0.05, ***p<0.01								

The internal structure is given by OLS regression and is based on a 5-point Likert scale. The OLS regression analysis shows insufficient results with an increased variance inflation factor (mean above 3). However, the relative positive influence of m4 and the negative influence of m6 and m7 can be derived as key points.

Overall, the industry of mechanical engineering faces a lack of trust towards lead users and is aware of basic costs and time constraints to employ this method as could be observed in this sample. This is in line with the 1st case study.

Table 62 OLS Regression Analysis in the Mechanical Engineering Sample

Strengths and Weaknesses of the lead User Method in Practice (n=19)	OLS Regression Coefficient (Std. Err.)	T-Value
M1: Lead users possess competencies that would enable a development of new solution concepts.	-0.62 (1.66)	-0.37
M2: Lead users have no strong interest that their own needs will be fulfilled.	-0.82 (1.49)	0.62
M3: Lead users are able to estimate product requirements in advance to ordinary customers.	0.88 (1.30)	0.67
M4: Lead user needs will reveal future needs of ordinary customers.	3.49 (2.13)	1.64
M5: Lead users provide ideas that are unknown yet.	-0.98 (1.26)	-0.78
M6: Lead user contributions are accepted by our engineers and developers.	-1.03 (0.88)	-1.17
M7: The Lead User Method is less costly in comparison to other methods of Innovation Management.	-1.49 (1.04)	-1.43
M8: The Lead User Method is less time consuming in comparison to other methods of Innovation Management.	1.09 (1.31)	0.83
M9: The trust to integrate lead users as the centre of innovation is given.	-0.25 (1.66)	-0.15
R-Square: 0.61; F-Value: 0.51; Significance: *p<=0.1, **p<0.05, ***p<0.01		

6.4. Verification in the German Field of Market Intelligence

6.4.1. The Application Field of Market Intelligence

The 3rd case study was done in the field of market intelligence, which covers the topic of *innovation management as a service*. This industry does consulting business with methodological usage of innovation tools for their clients. Innovation development projects cover a wide range of heterogeneous target markets w.r.t. their clients. Among these innovation management service agencies are specialised service providers that employ the lead user method, like the enterprises Hyve, "Lead Innovation Management", and earsandeyes. Their main objective is to develop innovations in collaboration with their clients in a serial manner.

The global *industry for market intelligence* generated a turnover of about US-$ 39 Billion in 2012 and US-$ 40.3 Billion in 2013 (see e.g. ESOMAR 2013 and ESOMAR 2014). About 40% of this revenue is generated in Europe and further 39% in North America. About 8.5% (US-$ 3.4 Billion) of the global turnover belongs to the German market (see ESOMAR 2013 for a more detailed analysis of the market for market intelligence). A significant 33% of the worldwide turnover in market intelligence is originated by the Top4 enterprises Nielsen Holding (USA), Kantar (GB), Ipsos (France) and GfK (Germany) (see GfK 2013, pp. 59-60 for a market overview). Today, the most important enterprises in Germany are GfK, TNS Infratest Holding, Nielsen Company, Ipsos Group, and Maritz Research in terms of the domestic generated turnover (ADM 2013). In sum, more than 15,300 consulting agencies exist in the German (BDU 2014).

The field of *innovation management covers about 7%* of the average turnover for the global market intelligence industry. This leads to a total turnover of US-$ 238 Million in the German market for innovation management and equals EUR 176 Million (exchange rate 1.355 as of June 10th 2014). The "Berufsverband Deutscher Markt- und Sozialforscher e.V." (BVM) is the most important lobby association for market and social research institutes in Germany (as given in a portrait by BVM 2014). The "Arbeitskreis Deutscher Markt- und Sozialforschungsinstitute e.V." (ADM) represents the interests of private-sector market and social research agencies in Germany, which in sum generate about 83% of the industry turnover (as given by data from ADM 2013).

Innovation service providers deal with *extreme usage contexts in terms of time constraints and financial restrictions* to perform development projects. Further, multiple customers and stakeholders, pre-defined innovation goals, and the variety of different application fields with heterogeneous needs, underlying trends, and dynamic operational lead user characteristics frame the basic environment for the implementation of the lead user method. Moreover, innovation service providers have to address the innovation project with limited resources and only gain restricted access to the clients' knowledge base and their processes. Personal interviews confirmed these findings. This background fosters an interesting basement for the evaluation of the statements.

Today, the lead user method costs about US-$ 75,000 (EUR 55,350) if it is applied in a business-to-business surrounding. This is based on Herstatt/von Hippel's (1992) estimation of the cost expenditure for concept generation and evaluation at the Hilti project. The projected costs include succeeding inflation rates of the German economy. This would lead to a maximal potential of about 3,200 lead user projects per year in Germany and about 36,000 global lead user projects per year performed by innovation service providers.

A preliminary analysis of the websites of BVM and ADM members was done to find sufficient evidence for the usage of the lead user method. This was completed by analysing offered methods on websites that imply the usage of lead users and by direct presentation of the "lead user" term on the website. It is known that the term "lead user" is also substituted by misleading terms like "pioneering user" or "extreme user" in reference to Lehnen et al. (2014). The examination of the members' websites was done using these analogue descriptions, too, and added terms like "creative workshops", "idea lab", and "concept garage". Overall, about 39% of the BVM and ADM members offer a (lead) user-based innovation management, with about 3.5% using a direct advertisement of the method.

6.4.2. Survey Preparation and Data Collection

The survey preparation started with five personal expert interviews to validate previous statements. The interviews have shown that the main motivation for the application of the lead user method is a *future oriented market intelligence*, since the retrospective view of traditional market intelligence was mentioned to be the main driver to resign from traditional approaches. The interviews also revealed that the main challenge is to *convince the client to open the innovation process* in reason of intellectual property law. When the client is convinced to do so, then the lead user method can achieve its full performance.

The methodological application is dynamically adapted to the external circumstances, like the addressed market or product category, and varies within the ranking and practical application of specific methodological phases. For example, the *identification process is simplified* by using data from a pre-defined *lead user panel* in multiple innovation projects. The expert interviews confirmed this finding at least twice and fostered this on several reasons like cost expenditures for lead user identification, time constraints, and trust as well as *familiarity with the user* to enable a successful collaboration. Other approaches rely on input by lead users from *analogous markets only* and fill the input from the target market by using the clients' product development staff. This input showed that the lead user method faces obstacles in terms of effectiveness, efficiency, NIH and a general doubt in relation with IP rights (see for example the ArbnErfG).

Thus, this study was reframed and used all of the previous statements. The survey itself was done using an adopted web-based questionnaire that consisted in sum of 21 interdependent questions separated into five basic groups. The introductory page of the questionnaire was redesigned and included a short movie to catch respondents' attention.

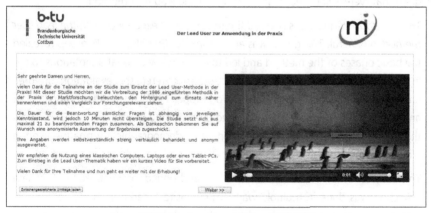

Figure 20 Introductory Page with Lead User Video

This was also used to visualise lead users along with a written description. This description was intended to *create an equal understanding of the term* "lead user". The introductory page is shown in figure 20. The textual description was on the left and the movie on the right.

The *1ˢᵗ group* of questions gathered basic information of the *respondent's knowledge of lead users and their general application*. The questionnaire was primarily dedicated to CEOs and project managers of the agencies. A *logical separation* of the questionnaire was used to cover (1) respondents being not familiar with lead users, (2) respondents being familiar with lead users but not using them, and (3) respondents being familiar with lead users and using them. If respondents were (1) not familiar with lead users, they were asked about other methods and tools from innovation management and their relevance (1... usage in no project; 5... usage in every project; 0... unknown). If respondents were (2) familiar with lead users, but do not employ them, they were asked for the reason of employing lead users. In this case, the statements were re-framed to express concerns towards lead user integration. If the respondent (3) employed lead users, the statements expressed benefits from employing lead users in the innovation process as presented in table 20. Further, in this case the 1ˢᵗ part of the questionnaire was extended to cover information about the respondents' clients, the usage intensity, the long-term trend in usage behaviour, and the innovation goals that were associated with the usage of the lead user method.

The *2ⁿᵈ group* of questions covered the *respondents' experience with the lead user method* in general. This group was applied in case (3) only. The *3ʳᵈ group* revealed the basic phases of the method and led to an overview of what adaptations to the lead user method were made. The *4ᵗʰ group* covered the evaluation of three presented approaches (in reference to von Hippel 1986, Herstatt/von Hippel 1992 and Sänn et al. 2013) to employ the lead user method in case (2) and (3). Overall, the *5ᵗʰ and final group* covered additional and demographic information like the industrial experience in years.

The data collection started in winter 2012 and ended in spring 2013. The data collection was done in multiple ways. The website "marktforschung.de" and its newsletter system were employed for online distribution. Personal acquisition of dedicated service providers and direct e-mailing to ADM and BVM members followed. This led to a 3rd round, which was as a reminder to take part in the survey.

6.4.3. Descriptive Results and Regression Analysis

The survey resulted in n=100 respondents that met the requirements. This leads to a general response rate of 26.6% in reference to ADM and BVM members with an average experience in market intelligence and innovation services of about 16 years. The application of the method was aimed at *product innovation* in 62.1% of all cases. This is in line with the generated turnover in market intelligence by the production industry (see ADM 2013). Table 63 provides an overview of the sample. About 65.5% of all applications were done in *business-to-consumer surroundings*.

Table 63 Descriptive in the Market Intelligence Sample

Descriptive Aspects	Sample Share
Amount of usable respondents	100 (26.6%)
Share of respondents being familiar with lead user	33 (33.0%)
Usage of the lead user method	15 (45.5% of experts)
Description of the Methodological Application (Multiple Answers Possible)	
Search for radical innovation	13.8%
Search for incremental innovation	34.5%
Appliance in service innovation	55.2%
Appliance in process innovation	10.3%
Appliance in product innovation	62.1%
Special appliance in consumer goods	65.5%
Special appliance in industrial goods	41.4%
Special appliance for product line optimisation	20.7%
Description of the Respondents' Clients (* clients as per EU Definition)	
Small-sized enterprises	27.6%
Medium-sized enterprises	62.1%
Greater enterprises	72.4%

The lead user method was familiar to 33 respondents. Further 45.5% of the respondents that were familiar with the method have already employed the method at least once. The method was used to generate *radical innovations* in about 65.5%. This is in line with basic literature. Further, 78.6% of the applications aimed at *complex innovations* and only 21.4% were intended to develop simple innovations. Complexity was defined as dealing with a high number of attributes, high innovation degree and sophisticated technical detailing.

The Shapiro-Wilks-Test for normal distribution shows significant ($p<0.01$) influences. The Mann-Whitney-U-test resulted in significant differences between users and non-users of the method. Table 64 shows the descriptive results in form of mean values per group and statement. The mean values indicate a general *uncertainty in the evaluation of the method* in contrast to the previous case studies.

Table 64 Results per Respondent Group in the Market Intelligence Sample

Strengths and Weaknesses of the Lead User Method in Practice	Respondents Being Familiar With the Lead User Method	
	Not applied n=18 Mean (Std. Dev.)	Applied n=15 Mean (Std. Dev.)
I1: Competencies that would enable a development of new solution concepts.	3.83 (1.10)	3.60 (1.06)
I2: Lead users have no strong interest that their own needs will be fulfilled.	3.00 (0.00)	2.67 (1.05)
I3: Estimation of product requirements in advance to ordinary customers.	3.11 (1.19)	4.07 (0.80)**
I4: Lead users are open-minded.	2.61 (0.70)	3.87 (1.06)***
I5: Lead user needs will reveal future needs of ordinary customers.	2.61 (0.78)	3.33 (1.05)*
I6: There are not enough innovative ideas available in our enterprise.	3.39 (1.04)**	2.67 (0.72)
I7: Lead users provide ideas that are unknown.	3.00 (0.00)	3.13 (1.13)
I8: Lead user contributions are accepted by our engineers and developers.	3.06 (0.24)	2.87 (0.92)
I9: The evaluation of lead user contributions is easy.	3.00 (0.00)***	2.33 (1.11)
I10: Lead user contributions fit our focus.	3.00 (0.00)	3.20 (0.68)
I11: Lead user contributions address a broad market.	2.33 (1.03)	3.13 (1.06)**
I12: The Lead User Method is less costly in comparison to other methods.	2.94 (1.51)	2.47 (0.74)
I13: The Lead User Method is less time consuming in comparison to other methods.	3.11 (1.28)*	2.20 (0.78)
I14: The trust to integrate lead users as the centre of innovation is given.	3.22 (1.06)	3.27 (0.96)
I15: The evaluation of market and technological trends is easy.	3.00 (0.00)***	2.87 (0.99)
I16: The Lead User Method faces broad support in the enterprise.	3.00 (0.00)	2.88 (0.96)
I17: The theoretical knowledge to implement the Lead User Method is available.	3.00 (0.77)	3.73 (1.16)*
I18: Fundamental IP rights and protective policies are from no concern.	3.11 (1.08)	2.91 (0.98)
Rating: 1 (totally disagree) ... 5 (totally agree); Significance: *p<=0.1, **p<0.05, ***p<0.01		

On the positive site, users confirm that lead users are open-minded for new ideas (i4), provide ideas that are unknown to the project team (i7), and provide contributions that fit in the focus of the company's innovation goal (i10). They also confirm that lead user *needs are able to reveal future needs of ordinary customers* (i5, p<0.01) and allow an early estimation of future product requirements (i3).

Moreover, lead user contributions will fit a broad market (i11). The latter one is seen negative (p<0.05) by non-users and points to positive results in the application of the method. Surprisingly, users and non-users agree that the trust to integrate lead users is given and that IP rights are of no concern (i14 and i18).

On the negative site, users confirm that lead users tend to have a drive to pursue their own needs (i2) and admit that *lead user contributions partially face NIH* (i8). This might be related to the fact that their contributions are hard to evaluate (i9) and that the evaluation of the underlying trends and technologies can be challenging (i15). The related *resource requirement* of the method is also seen as a negative aspect of the method (i12). Users point to extended time expenses to apply the method in a correct way (i13). This was rated low (mean value=2.20 out of 5 points) and was the worst rating in this survey.

Overall, the method partially *lacks an enterprise-wide support* (i16), although the fundamental theoretical knowledge to apply the method is available (i17). Surprisingly, users admit that there is a sufficient amount of innovative ideas available inside the company (i6), which shows significant difference (p<0.05).

A detailed look at the Pearson *correlations* between these statements is provided in table 67 with an inter-correlation matrix (at the end of this subchapter). In general, significant correlations could be observed.

Lead users' competencies to develop new solutions (i1) are in significant positive correlation with the acceptance by internal engineers and developers (i8, r=0.36, p<0.05) and with an easy evaluation of their contributions (i9, r=0.44, p<0.01). The higher the competencies, the better the acceptance and the easier the evaluation. This is aside with further positive correlations that point to a better fit to the innovation focus (i10, r=0.37, p<0.05) and an improved trust to integrate lead users (i14, r=0.47, p<0.01). Overall, *lead users' competencies favour the general acceptance of the method, but not necessarily favour the support*. However, a strong significant correlation between the evaluation of lead user contributions (i9) and available support within the company (i16, r=0.55, p<0.01) can be seen and is emphasised by an easy evaluation of trends (i15), which is in strong correlation with a broad support (i16, r=0.45, p<0.01).

Lead users' possibilities to estimate product requirements in advance to ordinary customers (i3) are a great benefit of the method. This is highly correlated with their open-mindedness (i4, r=0.49, p<0.01) and their ability to reveal future needs (i5, r=0.32, p<0.1), which will address a broader market (i11, r=0.32, p<0.1). It can be argued that companies are willing to spend more financial resources on the lead user method to benefit from this (i12, r=-0.30, p<0.1).

The lead users' open-mindedness (i4) correlates positive with the aforementioned ability to show future needs (i5, r=0.46, p<0.01), to generate new-to-the-company ideas (i7, r=0.34, p<0.1), and to achieve a broad market coverage (i11, r=0.42, p<0.05). Although, the correlation of i4 and the evaluation of underlying trends is negative and significant (i15, r=-0.54, p<0.01). Further positive and strong significant relations were seen between i5 and i7 (r=0.35, p<0.05), i7 and the internal acceptance (i8, r=0.60, p<0.01), and i7 and i10 (r=0.62, p<0.01). An overall internal acceptance (i8) is correlated with i9 (r=0.59, p<0.01), i12 (r=0.39, p<0.05), i13 (r=0.37, p<0.05), and shows trust (i14, r=0.45, p<0.01) along with support in the company (i16, r=0.67, p<0.01). A similar observation can be seen in the strong significant correlation between the focus of lead users' contributions (i10) and an addressed broad market (i11, r=0.49, p<0.01). In addition, the knowledge to implement the method (i17) is in positive correlation with aspects of patent rights (i18, r=0.35, p<0.05).

Negative aspects are drawn from the lead users' focus on own needs (i2). This *egocentric focus* is in significant negative correlation with their own open-mindedness (i4, r=-0.33, p<0.1) and raises doubt to predict future needs that will fit ordinary customers in a reliable way (i5, r=-0.79, p<0.01). This is further correlated with a negative ideation of new-to-the-company contributions (i7, r=-0.45, p<0.01), supports the existence of NIH (i8, r=-0.36, p<0.05), and points to the assumption that lead user contributions will not fit the focus (i10, r=-0.34, p<0.1). Further negative correlations can be seen between lead users' ability to reveal future needs (i5) and an easy evaluation of underlying trends (i15, r=-0.36, p<0.05). The strong tie between costs and time demands is given by the strong and significant correlation between i12 and i13 (r=0.69, p<0.01).

The internal structure that leads to the usage of the lead user method is given by OLS regression in accordance with previous analysis.

Table 65 OLS Regression Analysis in the Market Intelligence Sample

Strengths and Weaknesses of the Lead User Method in Practice	OLS Regression Coefficient (Std. Err.)	T-Value
I1: Competencies that would enable a development of new solution concepts.	-0.01 (0.23)	0.97
I2: Lead users have no strong interest that their own needs will be fulfilled.	-0.19 (0.52)	0.72
I3: Estimation of product requirements in advance to ordinary customers.	0.25 (0.22)	0.27
I4: Lead users are open-minded.	-0.05 (0.29)	-0.18
I5: Lead user needs will reveal future needs of ordinary customers.	0.10 (0.06)	0.25
I6: There are not enough innovative ideas available in our enterprise.	-0.28 (0.23)	-1.22
I7: Lead users provide ideas that are unknown.	0.23 (0.56)	0.41
I8: Lead user contributions are accepted by our engineers and developers.	0.35 (0.78)	0.44
I9: The evaluation of lead user contributions is easy.	0.40 (0.51)	0.78
I10: Lead user contributions fit our focus.	-0.19 (0.80)	-0.24
I11: Lead user contributions address a broad market.	0.49 (0.26)	1.90*
I12: The Lead User Method is less costly in comparison to other methods.	0.27 (0.26)	1.01
I13: The Lead User Method is less time consuming in comparison to other methods.	-0.70 (0.29)	-2.47**
I14: The trust to integrate lead users as the centre of innovation is given.	0.23 (0.25)	0.92
I15: The evaluation of market and technological trends is easy.	-1.21 (0.51)	-2.36**
I16: The Lead User Method faces broad support in the enterprise.	-0.44 (0.51)	-0.86
I17: The theoretical knowledge to implement the Lead User Method is available.	0.27 (0.26)	1.05
I18: Fundamental IP rights and protective policies are from no concern.	0.30 (0.31)	0.98
R-Square: 0.86^{***}; F-Value: 4.60; Significance: $^*p<=0.1$, $^{**}p<0.05$, $^{***}p<0.01$		

The OLS regression analysis shows sufficient results and reveals *time constraints* (i13) and the hard *evaluation of underlying trends* (i15) as major weaknesses that hinder the application of the lead user method.

The methodological application is further motivated by lead user input that will address a broader market (i11). This is confirmed by the previous rating of the statements (see table 64). The variance inflation factor (mean >4) demands an in-depth analysis using stepwise regression.

The resulting regression model indicates a sufficient model fit and reveals i4, i5, i13, and i17 as important aspects to decide whether to employ the lead user method or not. Another interpretation leads to the assumption that long-term innovation projects tend to rely on the lead user method. Lead users' open-mindedness (i4) and ability to predict future needs (i5) emphasise the methodological benefit. Table 66 illustrates the derived regression models.

Table 66 Stepwise OLS Regression in the Market Intelligence Sample

	OLS Regression Coefficient (Std. Err.)			
	Model 1	Model 2	Model 3	Model 4
I4	0.95 (0.20)***	0.87 (0.18)***	0.81 (0.17)***	0.62 (0.19)***
I5				0.43 (0.20)**
I13			-0.37 (0.16)**	-0.42 (0.15)**
I17		0.49 (0.19)**	0.52 (0.18)***	0.55 (0.17)***
	R-Square: 0.66***; F-Value: 23.94;	R-Square: 0.73***; F-Value: 17.45;	R-Square: 0.78***; F-Value: 15.12;	R-Square: 0.82***; F-Value: 13.83;
	Significance: *p<0.1, **p<0.05, ***p<0.01			

Table 67 Inter-Correlation Matrix in the Market Intelligence Sample

	(1)	(2)	(3)	(4)	(5)	(6)	(7)	(8)	(9)	(10)	(11)	(12)	(13)	(14)	(15)	(16)	(17)
(2)	-0.10	1.00															
(3)	0.18	-0.25	1.00														
(4)	0.04	-0.33*	0.49***	1.00													
(5)	0.07	-0.79***	0.32*	0.46***	1.00												
(6)	0.20	-0.12	-0.35**	-0.16	0.14	1.00											
(7)	0.22	-0.45***	0.15	0.34*	0.35**	0.30*	1.00										
(8)	0.36**	-0.36**	0.11	0.19	0.25	0.21	0.60***	1.00									
(9)	0.44***	0.03	-0.09	-0.26	-0.10	0.26	0.34*	0.59***	1.00								
(10)	0.37**	-0.34	0.20	0.28	0.29	0.27*	0.62***	0.22	0.33*	1.00							
(11)	0.03	0.06	0.31*	.417*	0.22	-0.04	0.29	-0.01	0.07	0.49***	1.00						
(12)	-0.13	-0.05	-0.30*	0.02	0.27	0.22	0.15	0.39**	0.29	-0.01	0.14	1.00					
(13)	-0.09	0.02	-0.18	-0.16	0.07	0.04	0.06	0.37**	0.20	-0.12	0.05	0.69***	1.00				
(14)	0.47***	0.01	0.05	0.02	-0.08	0.11	0.19	0.45***	0.32*	0.29	-0.10	0.03	0.01	1.00			
(15)	0.24	0.29	-0.23	-0.54***	-0.36**	0.18	-0.01	0.24	0.68***	-0.06	-0.06	0.28	0.32*	0.12	1.00		
(16)	0.29	-0.22	0.17	0.15	0.14	0.20	0.20	0.67***	0.55***	0.23	-0.03	0.13	0.10	0.45***	0.07	1.00	
(17)	0.00	-0.06	0.14	0.17	0.02	-0.21	-0.15	0.06	0.01	0.27	0.18	-0.02	0.04	0.19	-0.22	0.26	1.00
(18)	0.03	-0.02	-0.09	-0.04	0.03	-0.03	-0.07	0.33*	0.26	-0.05	-0.13	0.49***	0.65***	0.29	0.32	0.27	0.35**

Significance: *p<0.1, **p<0.05, ***p<0.01

6.5. Cross-Industry Findings

6.5.1. Methodological Basement

Revealed strengths and weaknesses in every industry point to the basic assumption that multiple aspects favour or hinder the usage of the lead user method within a company (as Lüthje/Herstatt 2004 already started to discuss). Lehnen et al. (2014) pointed to a missing understanding within industrial practice to employ the method. In detail, they named a missing understanding of scientific lead user characteristics, individually developed criteria to describe a lead user in practice, and adaptations in the identification process and the workshop setting. Furthermore, they could identify positive aspects of the method (e.g. risk reduction for product launch, customer orientation) as well as multiple weaknesses in its use (e.g. identification of lead users). A comparison of all industries may extend this.

6.5.2. Descriptive Results

Previous studies presented in chapters 6.2, 6.3, and 6.4 overlapped in selected questions for the evaluation of the lead user method. A *cross-industry comparison* is expected to reveal industry-specific influences on the evaluation task and to provide generalizable strengths and weaknesses. The overall studies generated n=311 respondents. The main target group of innovation managers, project managers, and innovation consultants was addressed and resulted in a share of 84.5% of all respondents. Surprisingly, only 27.1% of all respondents were familiar with the lead user method. This share was observable in every surveyed industry and ranged from 19.2% in the field of mechanical engineering to 33.0% in the field of market intelligence. Interestingly, a majority of about 82% of all respondents possesses more than five years of experience in their business.

The search for radical innovations in the three addressed industries is not the main driver to apply the lead user method. This is in line with experience from practice, which gives *incremental innovation* a significant role in today's innovation projects (see Cooper/Dreher 2010). In line with the basic scope of the lead user method is its *application for product innovation*. Moreover, the lead user method was also applied for *product line optimisation*. The results further show that in the field of market intelligence no application for process innovation is given.

In contrast, using lead users to develop *process innovations* seems to be known in the automotive industry and in the field of mechanical engineering (see table 68).

Table 68 Descriptive Summary of the Cross-Industry Sample

Description of the Sample	Autom. Ind.	Mech. Eng.	Market Intell.
Usable respondents and covered market share	112 (6.6%)	99 (2%)	100 (26.6%)
Respondents being familiar with lead users	32 (28.6%)	19 (19.2%)	33 (33.0%)
Usage of the lead user method	26 (81.3% of experts)	16 (84.2% of experts)	15 (45.5% of experts)
More than 5 years of experience in business	87,5%	79,8%	78,7%
Description of the methodological application (multiple answers possible)			
Search for radical innovation	37.5%	18.8%	6.7%
Search for incremental innovation	62.5%	81.2%	60.0%
Appliance in process innovation	25.0%	18.8%	0%
Appliance in product innovation	71.4%	81.2%	66.7%
Special appliance for product line optimisation	21.9%	50.0%	33.3%
Description of the respondents (* as per EU definition, clients' data for market intelligence)			
Small-sized enterprises	20.5%	29.3%	27.6%
Medium-sized enterprises	54.5%	42.4%	62.1%
Greater enterprises	25.0%	28.3%	72.4%

6.5.3. Strengths and Weaknesses

If the lead user method is known then it is rated as a *beneficial tool* to stimulate new problem solutions, innovation concepts, and ideas for future product components. This becomes clear in table 69. However, there are significant differences that could be observed and point to negative contingencies.

Lead users are said to have *no strong interest in solving their individual problems,* solely. This is widely accepted by users of the method, but shows significant differences ($p<0.05$) in comparison to non-users of the method. An even more important statement is that lead users' needs will *reveal future needs of ordinary customers*. Users of the method seem to emphasise this and rate this item with significant difference ($p<0.05$) in comparison to non-users of the method. The fact that lead users are able to *enhance in-house ideation* – to address a functional fixedness – and provide ideas that are unknown is rated significantly higher ($p<0.05$). Remarkably, users and non-users seem to agree equally on the facts that lead users are accepted by internal developers and engineers – to address the NIH – and that the method is less costly and time consuming.

Table 69 Cross-Industry Ratings of Methodological Benefits

Strengths and Weaknesses of the Lead User Method in Practice	Respondents Being Familiar with the Lead User Method	
	Not applied n=27 Mean (Std. Dev.)	Applied n=57 Mean (Std. Dev.)
Lead users possess competencies that would enable a development of new solution concepts.	3.74 (1.06)	3.77 (0.87)
Lead users have no strong interest that their own needs will be fulfilled exclusively.	3.52 (1.25)	4.13 (0.74)**
Lead user needs will reveal future needs of ordinary customers.	2.85 (0.91)	3.30 (0.89)**
Lead users provide ideas that are unknown yet.	3.22 (0.64)	3.54 (0.97)**
Lead user contributions are accepted by our engineers and developers.	3.19 (0.74)	3.36 (1.08)
The lead user method is less costly in comparison to other methods of innovation management.	2.89 (1.37)	3.09 (0.98)
The lead user method is less time consuming in comparison to other methods of innovation management.	2.85 (1.20)	2.89 (1.03)
The trust to integrate lead users as the centre of innovation is given.	2.93 (1.14)	2.89 (1.03)
Scale of Mean Value: 1 (totally disagree) ... 5 (totally agree); Significance: **$p < 0.05$		

Overall, the trust to open the innovation process for lead users is given. This is in contrast to literature and the derived weaknesses of the method in previous findings. Especially the trust to integrate lead users is in doubt.

Table 70 highlights these questions and performs a separated *industry-specific analysis*. The ratings among all users of the method significantly differ between each industry (p<0.01). Lead users' contributions are well accepted in the field of mechanical engineering with a mean rating of 4.25 out of 5.

In contrast, the *field of market intelligence* seems to face challenges in reference to the acceptance by internal engineers and developers. This can especially be the case for their clients. Market intelligence typically uses methods of innovation management as a consulting service. Thus, from a respondent's view this question can point to the acceptance by their customers' internal development staff. Further, the lead user method is not seen as being less costly and less time consuming as literature (e.g. Herstatt/von Hippel 1992) indicates. The ratings differ significantly to the users in other industries. This seems to be the case since the lead user method requires a custom application for each application field and thus requires human resources and time.

The trust to integrate lead users as the centre of innovation is not given in the *field of mechanical engineering*. All users interviewed pointed to this fact, which may be fostered by the industry's structure with a majority of SMEs. The basic problem of product counterfeiting and piracy are major challenges for SMEs. This holds especially true in the field of mechanical engineering. In addition, products in this industry are also customer specific and thus one might raise the question why lead users should be incorporated if there is already a close relation between internal developers and the customer in NPD.

Table 70 Segmented Results for Methodological Benefits in all Industries

Strengths and Weaknesses of the Lead User Method in Practice	Users of the Method per Industry		
	Automotive Industry n=26 Mean (Std. Dev.)	Mechanical Engineering n=16 Mean (Std. Dev.)	Market Intelligence n=15 Mean (Std. Dev.)
Lead user contributions are accepted by our engineers and developers.	3.23 (0.99)	4.25 (0.97)***	2.87 (0.92)
The lead user method is less costly in comparison to other methods of innovation management.	3.12 (0.91)	3.67 (0.98)***	2.47 (0.74)
The lead user method is less time consuming in comparison to other methods of innovation management.	2.88 (0.91)	3.56 (1.03)***	2.20 (0.78)
The trust to integrate lead users as the centre of innovation is given.	3.35 (0.80)***	1.81 (0.54)	3.27 (0.96)
Scale of Mean Value: 1 (totally disagree) ... 5 (totally agree); Significance: ***$p<0.01$			

The previous surveys incorporated a ranking of the traditional lead user method by Eric von Hippel from 1986, the Hilti approach by Cornelius Herstatt and Eric von Hippel from 1992, and the Preference-Driven Lead User Method. Respondents who were familiar with the lead user method were asked to choose the best and the worst approach among these three options in reference to their demand.

The three methods were presented as a process plan with their basic phases, but without any reference or name. The new method was shrunken to fit the typical 4-phase layout to guarantee unbiased comparability.

Table 71 4-Phase Layout of 3 Presented Lead User Methods for Evaluation

	von Hippel (1986)	Herstatt/von Hippel (1992)	The New Method
Phase 1	Identification of market- and technology trends	Identification of needs and trends	Identification of needs and trends
Phase 2	Identification of lead users	Identification of lead users	Identification of lead users
Phase 3	Analysis of lead user needs (and ideas)	Development of a concept via workshop setting	Comparison of lead user contributions to the overall market
Phase 4	Applying lead user data to target market	Testing concepts appeal to ordinary users	Lead user workshop

Overall, the Preference-Driven Lead User Method was chosen as the most beneficial approach to be implemented in 18.5% of all cases (n=81). The *Hilti approach was mentioned as the best method* in 61.7% of all cases. The traditional method was mentioned in 19.8% of all cases. In contrast, the traditional approach was mentioned to be the worst option of all three in 53.1% of all cases. The Preference-Driven Lead User Method was mentioned to be the worst option in 40.7% and was chosen as the *2nd best option* in 40.7% of all cases. A strong, negative and significant correlation could be observed between favouring the Preference-Driven Lead User Method and the rejection of the traditional method (r=-0.68, p<0.01).

A further separation of *respondents with 5+ years of experience* in their markets shows interesting changes. If this filter is applied then the Preference-Driven Lead User Method becomes more interesting and is favoured in 20.9% of all cases. The rejection of the new method drops slightly from 40.7% to 37.3%. The new method becomes the 2nd best option in 41.8% of all cases.

A further separation using the *usage intensity for the lead user method* showed no significant correlations, but revealed that users with low usage intensity favour the new approach (60% of all favoured cases). Further, a *separation per industry* shows that market intelligence and the automotive industry favours the new method over the traditional one on average.

6.6. Implications

The survey of strengths and weaknesses in the three application fields analysed n=84 respondents that were familiar with the lead user method. The analysis shows that the lead user method is known by 25% of all respondents. If the method is known, then its usage could be observed quite often (71%). The field of innovation service applied the lead user method to develop radical innovations (66%) in contrast to the automotive industry and mechanical engineering that generate incremental innovations (72%). The major focus is on product innovation in all three samples. The automotive industry is mainly driven by mass production whereas the mechanical engineering industry is mainly driven by customer-based development orders. The field of innovation services covers clients from different industries and addresses business-to-consumer surroundings in 65.5% of all cases.

The analysis on mean values has revealed time and cost constraints, the trust in the method, and the NIH as major impacts to hinder the application of the lead user method. The last two might be triggered by an egocentric focus on lead users' individual needs. Thus, the evaluation of their contributions and the underlying trends are affected. If they are not egocentric, then lead users are able to predict future product requirements and ordinary customers' needs. This focus and the ability to derive implications for ordinary customers are in line with the theoretical model. In addition, lead users' competences to develop own solutions are a major impact that favours the usage of the method and builds significant benefits for practitioners. The correlation analysis and the significant differences between users and non-users of the method reveal that the evaluation of the underlying trends is also of major concern. Surprisingly, Intellectual Property Rights have no major influence on the usage of lead users.

Overall, the method is rarely used in practice (mean value=2.3) unlike more traditional methods like focus groups (mean value=3.33, $p<0.05$; see Cooper/Dreher 2010 for confirmation). This might be related with an unwillingness to open the innovation process and thus a lack of use experience could be observed. However, previous findings of strengths and weaknesses of the lead user method in theory were proven.

7. Conclusion and Outlook

7.1. Conclusion

The present thesis proposed the Preference-Driven Lead User Method for NPD.

The thesis developed the necessity of the new method by analysing and reflecting the traditional lead user method. The argumentation highlighted open questions and weaknesses of the method that are derived from findings in literature. Thus, multiple aspects of the traditional lead user method remained questionable for practice when dealing with *restricted resources*.

Overall, the traditional lead user method deals with the benefit that lead users are ahead of the market. Thus, their contributions address needs and demands that may never be experienced by ordinary customers. Literature points to the fact that the acceptance of lead user contributions is questionable and thus a diffusion process may be postponed. This becomes critical for e.g. SMEs that deal with restricted resources for NPD. In a sequential NPD process, the successive preference measurement will provide information about the preferability of the new concept.

However, *ordinary customers are not reliable* in expressing their needs and demands. Thus, the argumentation to favour lead users for concept development can be applied to argue against preference measurement. This becomes evidence e.g. in high-technology markets as findings in literature emphasise. As a result, the *concept generation fails* and additional resources are required to perform another development. Also, the project team may *neglect promising user contributions*.

This leads to the research question *"Can the lead user method and preference measurement be combined to result in an integrated method for new product development?"*.

The traditional lead user method and preference measurement are combined to address this question. The proposed integrated method connects idea generation and evaluation using collaborative filtering to determine the similarity among respondents and predict their preferences on (lead) user contributions. This method is exemplary illustrated in mountain biking.

The method is further empirically tested in a complex high-technology market – the market for IIT security. The results show that the outcome of the Preference-Driven Lead User Method was favoured in the term of market potential.

The *applicability of the new method was further analysed within the markets of the German automotive industry, mechanical engineering, and market intelligence.* The results point to multiple aspects that were criticised in relation to the traditional lead user method and even led to a rejection on implementing the traditional method in practice. Overall, the new method was compared with two standard lead user methods for implementation in these three application fields. The results showed that the new Preference-Driven Lead User Method was the 2nd best option.

A *discussion* of the new method can argue about the demand for data quality since collaborative filtering was performed using all respondents. It is the case that users contribute and reveal their thoughts and needs to the user community, but fail to answer standard preference measurement in a reliable way. Although, their contributions were kept in the survey, revealed to the user community, and accepted by a majority of the respondents. Their own preferences were skipped in the analysis that is caused by the methodological background of quantitative preference measurement.

The Preference-Driven Lead User Method is further applicable for in-house lead user workshops, e.g. when in-house suggestion systems are gathering a high amount of contributions. In addition, the project team can focus on process innovation to realise lead user contributions that are proven to be accepted within the market, but do not fit the given resources (after the method). Further recommendations for application in practice suggest a gamification of the contribution and measurement parts. Also, users of the Preference-Driven Lead User Method need to take the addressed community into account. The results from this method may be improved by developing a dedicated community. West/Lakhani (2008) point to the relevance of nursing and developing communities by companies, but warn about not contributing resources back to the community. The negative effects of hijacking user (community) innovations are mentioned in subchapter 2.2.5 and were noted by e.g. Raasch et al. 2008 and Franke et al. 2013.

7.2. Outlook

The *improvement of the Preference-Driven Lead User Method* is in the focus for future research. Thus, the new method needs to be separated to its components to highlight open research questions.

The *classification of lead users* remains one of the most important aspects for the lead user component. It is not necessarily important for the classification of user contributions, but to determine similar or different evaluation structures of a lead user group and a non-lead user group.

Along with this is the discussion of the *motivational background* of lead users and how to improve free revealing and participation. Modern approaches of gamification can support the willingness to participate. Advanced approaches can make use of netnography and porter stemming to identify lead users online. Further, it remains an important decision what model of *preference measurement* to choose, e.g. hybrid methods like WBUM. This decision is highly dependent on the complexity and the category of a product. An implementation using a maxdiff-approach to measure preferences seems to be promising since the cognitive effort of the respondent is expected to decrease. Thus, required cognitive capacity may be available to perform evaluation tasks with complex attributes.

User-based collaborative filtering is employed in the prototype of the Preference-Driven Lead User Method since the basic assumption fits the problem task. One could see that this is restricted in reason of binary data. Further, the calculation of users' similarity may be fostered on additional variable from lead user classification. The applicability can then be proven by preference measurement to determine equal preference structures within the lead user group. Now, UBCF makes use of similar rating structures per respondent. This becomes challenging when assuming that lead users and non-lead users will significantly differ in their ratings. Thus, collaborative filtering may be applied via profile matching (see Krzywicki et al. 2015). Overall, the accuracy of collaborative filtering may be improved by using multiple imputations (see Göthlich 2009, p. 128-129 in reference to Rubin 1977).

However, this study is not without restrictions. Thus, research needs to prove the applicability and additional advantages in further empirical studies.

Literature

Achilladelis, B., P. Jervis, and A. Robertson. 1971. Project SAPPHO: A Study of Success and Failure in Industrial Innovation. *2 vols. Center for the Study of Industrial Innovation,* London.

Adamson, R. E. 1952. Functional Fixedness as Related to Problem Solving: A Repetition of Three Experiments. *Journal of Experimental Psychology* 44 (4): 288-291.

Adamson, R. E. and D. W. Taylor. 1954. Functional Fixedness as Related to Elapsed Time and to Set. *Journal of Experimental Psychology* 47 (2): 122-126.

Addelman, S. 1962. Orthogonal Main-effect Plans for Asymmetrical Factorial Experiments. *Technometrics* 4 (1): 21-46.

ADM. 2013. *ADM Jahresbericht 2012.* Frankfurt am Main: ADM Arbeitskreis Deutscher Markt- und Sozialforschungsinstitute e. V.

Ahtola, O. T. 1975. The Vector Model of Preferences: An Alternative to the Fishbein Model. *Journal of Marketing Research* 12 (1): 52-59.

Akaah, I. P. and P. K. Korgaonkar. 1983. An Empirical Comparison of the Predictive Validity of Self-Explicated, Huber-Hybrid, Traditional Conjoint, and Hybrid Conjoint Models. *Journal of Marketing Research* 20 (2): 187-197.

Allen, T. J. and D. G. Marquis. 1964. Positive and Negative Biasing Sets: The Effects of Prior Experience on Research Performance. *IEEE Transactions on Engineering Management* 11 (4): 158-161.

Altmann, G. 2003. *Unternehmensführung und Innovationserfolg: Eine empirische Untersuchung im Maschinenbau.* Wiesbaden: Deutscher Universitäts-Verlag.

Al-Zu'bi, Z. M. F. and C. Tsinopoulos. 2012. Suppliers versus Lead Users: Examining their Relative Impact on Product Variety. *Journal of Product Innovation Management* 29 (4): 667-680.

Al-Zu'bi, Z. M. F. and C. Tsinopoulos. 2013. An Outsourcing Model for Lead Users: An Empirical Investigation. *Production Planning & Control* 24 (4-5): 337-346.

Amazon. 2014a. *Annual Report 2013.* Seattle: amazon.com, Inc.

Antons, D. and F. Piller. 2014. Opening the Black Box of" Not-Invented-Here": Attitudes, Decision Biases, and Behavioral Consequences. *The Academy of*

Management Perspectives. Available from http://dx.doi.org/10.5465/amp.2013.0091.

Archibugi, D. and S. Iammarino. 1999. The Policy Implications of the Globalisation of Innovation. *Research Policy* 28 (2-3): 317-336.

August, T. and T. I. Tunca. 2011. Who Should Be Responsible for Software Security? A Comparative Analysis of Liability Policies in Network Environments. *Management Science* 57 (5): 934-959.

Auty, S. 1995. Using Conjoint Analysis in Industrial Marketing: The Role of Judgment. *Industrial Marketing Management* 24 (3): 191-206.

Baier, D. 1999. Methoden der Conjointanalyse in der Marktforschungs- und Marketingpraxis. In *Mathematische Methoden der Wirtschaftswissenschaften*, ed. D. Baier, W. Gaul, and M. Schader, 197-206. Heidelberg: Physica.

Baier, D. 2000. *Marktorientierte Gestaltung innovativer Produkte und Dienstleistungen*. Habilitation. Universität Karlsruhe.

Baier, D. and M. Brusch. 2009. *Conjointanalyse: Methoden-Anwendungen-Praxisbeispiele*. Berlin: Springer-Verlag.

Baier, D. and A. Sänn. 2013. Lead User bei der Entwicklung neuer Produkte. *wisu - Das Wirtschaftsstudium* 42 (6): 799-804.

Baier, D. and A. Sänn. 2015. Marktforschung auf Industriegütermärkten. In *Handbuch Business-to-Business Marketing*, ed. K. Backhaus, and M. Voeth, 73-99. Wiesbaden: Springer Fachmedien.

Balcet, G., H. Wang, and X. Richet. 2012. Geely: A Trajectory of Catching up and Asset–Seeking Multinational Growth. *International Journal of Automotive Technology and Management* 12 (4): 360-375.

Baldwin, C. and E. von Hippel. 2011. Modeling a Paradigm Shift: From Producer Innovation to User and Open Collaborative Innovation. *Organization Science* 22 (6): 1399-1417.

Baltrunas, L., T. Makcinskas, and F. Ricci. 2010. Group Recommendations with Rank Aggregation and Collaborative Filtering. In *Proceedings of the 4th ACM*

Conference on Recommender Systems. Barcelona, Spain, September 26-30, 2010.

Barczak, G., A. Griffin, and K. B. Kahn. 2009. Perspective: Trends and Drivers of Success in NPD Practices: Results of the 2003 PDMA Best Practices Study. *Journal of Product Innovation Management* 26 (1): 3-23.

Bass, F. M. 1969. A New Product Growth for Model Consumer Durables. *Management Science* 15 (5): 215-227.

Bass, F. M. 2004. Comments on "A New Product Growth for Model Consumer Durables": The Bass Model. *Management Science* 50 (12): 1833-1840.

Bass, F. M., T. V. Krishnan, and D. C. Jain. 1994. Why the Bass Model Fits without Decision Variables. *Marketing Science* 13 (3): 203-223.

Batinic, B., C. M. Haupt, and J. Wieselhuber. 2006. Validierung und Normierung des Fragebogens zur Erfassung von Trendsetting (TDS). *Diagnostica* 52 (2): 60-72.

BDU 2014: *Facts & Figures zum Beratermarkt 2013/2014*. Bonn: Bundesverband Deutscher Unternehmensberater BDU e.V.

Behrens, J. and H. Ernst. 2014. What Keeps Managers Away from a Losing Course of Action? Go/Stop Decisions in New Product Development. *Journal of Product Innovation Management* 31 (2): 361-374.

Belz, F.-M. and W. Baumbach. 2010. Netnography as a Method of Lead User Identification. *Creativity and Innovation Management* 19 (3): 304-313.

Benjamin, V. and H. Chen. 2012. Securing Cyberspace: Identifying Key Actors in Hacker Communities. In *2012 IEEE International Conference on Intelligence and Security Informatics (ISI)*. Washington, D. C., USA, June 11-14, 2012.

Bensch, R., S. Selka, and A. Sänn. 2012. How Facebook Influences Validity in Online-Surveys. In *34th INFORMS Marketing Science Conference*. Boston, USA, June 7-9, 2012.

Bichler, A. and V. Trommsdorff. 2009. Präferenzmodelle bei der Conjointanalyse. In *Conjointanalyse*, ed. D. Baier, and M. Brusch, 59-71. Berlin: Springer-Verlag.

Bigler, L. and R. Drenth. 2013. Die neue Rolle des Marketing im Buying Center bei industriellen ICT-Investitionen. *Marketing Review St. Gallen* 30 (4): 36-51.

Bilgram, V., A. Brem, and K.-I. Voigt. 2008. User- Centric Innovations in New Product Development - Systematic Identification of Lead Users Harnessing Interactive and Collaborative Online-Tools. *International Journal of Innovation Management* 12 (3): 419-458.

Birch, H. G. and H. S. Rabinowitz. 1951. The Negative Effect of Previous Experience on Productive Thinking. *Journal of Experimental Psychology* 41 (2): 121-125.

BITKOM. 2008. *Studie zur Bedeutung des Sektors Embedded-Systeme in Deutschland.* Berlin: BITKOM Bundesverband Informationswirtschaft, Telekommunikation und neue Medien e. V.

BITKOM. 2010. *Eingebettete Systeme – Ein strategisches Wachstumsfeld für Deutschland.* Berlin: BITKOM Bundesverband Informationswirtschaft, Telekommunikation und neue Medien e. V.

Bjelland, O. M., and R. C. Wood. 2008. An Inside View of IBM's 'Innovation Jam'. *MIT Sloan Management Review* 50 (1): 32-40.

BMI. 2009. *Nationale Strategie zum Schutz Kritischer Infrastrukturen (KRITIS-Strategie).* Berlin: BMI Bundesministerium des Innern.

Böcker, F. 1986. Präferenzforschung als Mittel marktorientierter Unternehmensführung. *Zeitschrift für betriebswirtschaftliche Forschung* 38 (7/8): 543-574.

Bogers, M., A. Afuah, and B. Bastian. 2010. Users as Innovators: A Review, Critique, and Future Research Directions. *Journal of Management* 36 (4): 857-875.

Bonebright, D. A. 2010. 40 Years of Storming: A Historical Review of Tuckman's Model of Small Group Development. *Human Resource Development International* 13 (1): 111-120.

Brabham, D. C. 2010. Moving the Crowd at Threadless: Motivations for Marticipation in a Crowdsourcing Application. *Information, Communication & Society* 13 (8): 1122-1145.

Brabham, D. C. 2012. Motivations for Participation in a Crowdsourcing Application to Improve Public Engagement in Transit Planning. *Journal of Applied Communication Research* 40 (3): 307-328.

Bright, J. R. 1969. Some Management Lessons from Technological Innovation Research. *Long Range Planning* 2 (1): 36-41.

Bruns, H. C. 2013. Working Alone Together: Coordination in Collaboration across Domains of Expertise. *Academy of Management Journal* 56 (1): 62-83.

Brusch, M. 2005. *Präferenzanalyse für Dienstleistungsinnovationen mittels multimedialgestützter Conjointanalyse*. Wiesbaden: Deutscher Universitäts-Verlag.

Buck, A., C. Herrmann, and D. Lubkowitz. 1998. *Handbuch Trendmanagement: Innovation und Ästhetik als Grundlage unternehmerischer Erfolge*. Frankfurt am Main: Frankfurter Allgemeine Zeitung Verlagsbereich Buch.

Butt, F. M. 2014. Effects of involving Lead Users and Close Customers in enhancing the process of New Service Development: a Study on Banking Sector. *IOSR Journal of Business and Management* 16 (1): 98-103.

BVM. 2014. *Verbandsprofil*. Available from http://bvm.org/bvmprofil/ [accessed 31 May 2014].

Cacheda, F., V. Carneiro, D. Fernández, and V. Formoso. 2011. Comparison of Collaborative Filtering Algorithms: Limitations of Current Techniques and Proposals for Scalable, High-Performance Recommender Systems. *ACM Transactions on the Web* 5 (1): 1-33.

Carbonell, P., A. I. Rodriguez-Escudero, and D. Pujari. 2012. Performance Effects of Involving Lead User and Close Customers in New Service Development. *Journal of Services Marketing* 26 (7): 497-509.

Cattin, P. and D. R. Wittink. 1982. Commercial Use of Conjoint Analysis: A Survey. *Journal of Marketing* 46 (3): 44-53.

Chen, Y.-C., R.-A. Shang, and C.-Y. Kao. 2009. The Effects of Information Overload on Consumers' Subjective State Towards Buying Decision in the Internet Shopping Environment. *Electronic Commerce Research and Applications* 8 (1): 48-58.

Chesbrough, H. W. 2003. The Era of Open Innovation. *MIT Sloan Management Review* 44 (3): 35-41.

Chesbrough, H. and S. Brunswicker. 2014. A Fad or a Phenomenon?: The Adoption of Open Innovation Practices in Large Firms. *Research-Technology Management* 57 (2): 16-25.

Christensen, C. M., and J. L. Bower. 1996. Customer Power, Strategic Investment, and the Failure of Leading Firms. *Strategic Management Journal* 17 (3): 197-218.

Christensen, C. and J. Euchner. 2011. Managing Disruption: An Interview with Clayton Christensen. *Research Technology Management* 54 (1): 11-17.

Chrzan, K. and N. Golovashkina. 2006. An Empirical Test of Six Stated Importance Measures. *International Journal of Market Research* 48 (6): 717-740.

Cohen, J. B., M. Fishbein, and O. T. Ahtola. 1972. The Nature and Uses of Expectancy-Value Models in Consumer Attitude Research. *Journal of Marketing Research* 9 (4): 456-460.

Cooper, R. G. 2011. Perspective: The Innovation Dilemma: How to Innovate When the Market Is Mature. *Journal of Product Innovation Management* 28 (S1): 2-27.

Cooper, R. G. and A. Dreher. 2010. Voice-of-Customer Methods: What is the Best Source of New-Product Ideas? *Marketing Management* 19 (4): 38-43.

Cooper, R. G. and S. Edgett. 2008. Ideation for Product Innovation: What are the Best Methods. *PDMA visions magazine* 1 (1): 12-17.

Cooper, R. G. and E. J. Kleinschmidt. 1987. New Products: What Separates Winners from Losers? *Journal of Product Innovation Management* 4 (3): 169-184.

Cooper, R. G. and E. J. Kleinschmidt. 2000. New Product Performance: What Distinguishes the Star Products. *Australian Journal of Management* 25 (1): 17-46.

Cowen, A. P. 2012. An Expanded Model of Status Dynamics: The Effects of Status Transfer and Interfirm Coordination. *Academy of Management Journal* 55 (5): 1169-1186.

Creusen, M., E. J. Hultink, and K. Eling. 2013. Choice of Consumer Research Methods in the Front End of New Product Development. *International Journal of Market Research* 55 (1): 81-104.

Cvijikj, I. P. and F. Michahelles. 2011. The Toolkit Approach for End-user Participation in the Internet of Things. In *Architecting the Internet of Things*, ed. D. Uckelmann, M. Harrison, and F. Michahelles, 65-96. Berlin: Springer-Verlag.

Dahl, D. W. and P. Moreau. 2002. The Influence and Value of Analogical Thinking During New Product Ideation. *Journal of Marketing Research* 39 (1): 47-60.

Dahlander, L. and M. Magnusson. 2008. How do Firms Make Use of Open Source Communities? *Long Range Planning* 41 (6): 629-649.

Dannenberg, J. 2005. Von der Technik zum Kunden. In *Markenmanagement in der Automobilindustrie. Die Erfolgsstrategien internationaler Top-Manager*, ed. B. Gottschalk, R. Kalmbach, and J. Dannenberg, 33-58. Wiesbaden: Gabler.

de Jong, J. 2014. The Empirical Scope of User Innovation. In *Revolutionizing Innovation: Users, Communities, and Open Innovation*, ed. K. Lakhani, and D. Harhoff, forthcoming. Cambridge: The MIT Press.

de Jong, J. and O. Marsili. 2006. The Fruit Flies of Innovation: A Taxonomy of Innovative Small Firms. *Research Policy* 35 (2): 213-229.

de Jong, J. and E. von Hippel. 2009. Transfers of User Process Innovations to Process Equipment Producers: A Study of Dutch High-Tech Firms. *Research Policy* 38 (7): 1181-1191.

de Jong, J., E. von Hippel, F. Gault, J. Kuusisto, and C. Raasch. 2014. Market failure in the Diffusion of Consumer-Developed Innovations: Patterns in Finland. *Working Paper*. RSM Erasmus University.

Debruyne, M., R. Moenaert, A. Griffin, S. Hart, E. J. Hultink, and H. Robben. 2002. The Impact of New Product Launch Strategies on Competitive Reaction in Industrial Markets. *Journal of Product Innovation Management* 19 (2): 159-170.

Deci, E. L. and R. M. Ryan. 1993. Die Selbstbestimmungstheorie der Motivation und ihre Bedeutung für die Pädagogik. *Zeitschrift für Pädagogik* 39 (2): 223-238.

Decker, R. and R. Wagner. 2008. Fehlende Werte: Ursachen, Konsequenzen und Behandlung. In *Handbuch Marktforschung*, ed. A. Hermann, C. Homburg, and M. Klarmann, 53-80. Wiesbaden: Gabler.

DeMarco, T. 1979. *Structured Analysis and System Specification.* New Jersey: Prentice Hall.

DeMartino, R., H. M. Neck, P. A. Dwyer, and C. Treese. 2012. Radical Innovation in Medium-Sized Enterprises: An Exploratory Study of Strategic Intent and Innovation Process. *International Journal of Entrepreneurship and Innovation Management* 15 (3): 216-234.

DeSarbo, W. S., D. Fong, J. Liechty, and J. Coupland. 2005. Evolutionary Preference/Utility Functions: A Dynamic Perspective. *Psychometrika* 70 (1): 179-202.

DeSarbo, W. S., V. Ramaswamy, and S. H. Cohen. 1995. Market Segmentation with Choice-Based Conjoint Analysis. *Marketing Letters* 6 (2): 137-147.

Destatis. 2013. *Statistisches Jahrbuch: Deutschland und Internationales.* Wiesbaden: Statistisches Bundesamt.

Dhar, R. and K. Wertenbroch. 2000. Consumer Choice between Hedonic and Utilitarian Goods. *Journal of Marketing Research* 37 (1): 60-71.

Droge, C., M. A. Stanko, and W. A. Pollitte. 2010. Lead Users and Early Adopters on the Web: The Role of New Technology Product Blogs. *Journal of Product Innovation Management* 27 (1): 66-82.

Duncker, K. 1945. On Problem-Solving. *Psychological Monographs* 58 (5): 1-113.

Eckert, C. 2008. *IT-Sicherheit* 5th ed .München: Oldenbourgh Wissenschaftsverlag.

Eckert, J. and R. Schaaf. 2009. Verfahren zur Präferenzmessung – Eine Übersicht und Beurteilung existierender und möglicher neuer Self-Explicated-Verfahren. *Journal für Betriebswirtschaft* 59 (1): 31-56.

Edl, B, N. Franke, and E. von Hippel. 2014. Do People Encounter New Need-Solution Pairs? Evidence from Living & Interieur Fair Suggests: YES. Boston: Open and User Innovation Workshop. Available upon request.

Edwards, W. 1954. The Theory of Decision Making. *Psychological Bulletin* 51 (4): 380-417.

Eggers, F. and F. Eggers. 2011. Where Have All the Flowers Gone? Forecasting Green Trends in the Automobile Industry with a Choice-Based Conjoint Adoption Model. *Technological Forecasting and Social Change* 78 (1): 51-62.

Eisenberg, I. 2011. Lead-User Research for Breakthrough Innovation. *Research Technology Management* 54 (1): 50-58.

Eisenhardt, K. M., and J. A. Martin. 2000. Dynamic Capabilities: What Are They? *Strategic Management Journal* 21 (10/11): 1105-1121.

Elger, J. and C. Haußner. 2010. Entwicklungen in der Automatisierungstechnik. In *Internet der Dinge in der Intralogistik*, ed. W. Günthner, and M. ten Hompel, 23-27. Berlin: Springer-Verlag.

Elrod, T., R. D. Johnson, and J. White. 2004. A New Integrated Model of Noncompensatory and Compensatory Decision Strategies. *Organizational Behavior and Human Decision Processes* 95 (1): 1-19.

Elsberg, M. 2012. *Blackout: Morgen ist es zu spät*. München: Blanvalet-Verlag.

Enkel, E. and O. Gassmann. 2010. Creative Imitation: Exploring the Case of Cross-Industry Innovation. *R&D Management* 40 (3): 256-270.

Enkel, E., J. Perez-Freije, and O. Gassmann. 2005. Minimizing Market Risks Through Customer Integration in New Product Development: Learning from Bad Practice. *Creativity and Innovation Management* 14 (4): 425-437.

ESOMAR. 2013. *Global Market Research 2013*. Amsterdam: ESOMAR.

European Commission. 2005. *Die neue KMU-Definition: Benutzerhandbuch und Mustererklärung*. Brussels: European Commission.

Faullant, R., E. J. Schwarz, I. Krajger, and R. J. Breitenecker. 2012. Towards a Comprehensive Understanding of Lead Userness: The Search for Individual Creativity. *Creativity and Innovation Management* 21 (1): 76-92.

Feldman, L. P. and G. M. Armstrong. 1975. Identifying Buyers of a Major Automotive Innovation. *Journal of Marketing* 39 (1): 47-53.

Feng, Q., Z. Wang, D. Gu, and Y. Zeng. 2011. Household Vehicle Consumption Forecasts in the United States, 2000 to 2025. *International Journal of Market Research* 53 (5): 593-618.

Fidler, D. P. 2011. Was Stuxnet an Act of War? Decoding a Cyberattack. *IEEE Computer and Reliability Societies* 9 (4): 56-59.

Fishbein, M. 1967. Attitude and the prediction of behavior. In *Reading in Attitude Theory and Measurement*, ed. M. Fishbein, 477-492. New York: Wiley.

Fixson, S. K., Y. Ro, and J. K. Liker. 2005. Modularisation and Outsourcing: Who Drives Whom? A Study of Generational Sequences in the US Automotive Cockpit Industry. *International Journal of Automotive Technology and Management* 5 (2): 166-183.

Foxall, G. and J. Tierney. 1984. From CAP 1 to CAP 2: User-Initiated Innovation from the User's Point of View. *Management Decision* 22 (5): 3-15.

Franke, N. and E. von Hippel. 2003. Satisfying Heterogeneous User Needs via Innovation Toolkits: The Case of Apache Security Software. *Research Policy* 32 (7): 1199-1215.

Franke, N. and S. Shah. 2003. How Communities Support Innovative Activities: An Exploration of Assistance and Sharing Among End-Users. *Research Policy* 32 (1): 157-178.

Franke, N., E. von Hippel, and M. Schreier. 2006. Finding Commercially Attractive User Innovations: A Test of Lead User Theory. *Journal of Product Innovation Management* 23 (4): 301-315.

Franke, N., M. Schreier, and U. Kaiser. 2010. The "I Designed It Myself" Effect in Mass Customization. *Management Science* 56 (1): 125-140.

Franke, N., P. Keinz, and K. Klausberger. 2013. "Does This Sound Like a Fair Deal?": Antecedents and Consequences of Fairness Expectations in the Individual's Decision to Participate in Firm Innovation. *Organization Science* 24 (5): 1495-1516.

Franke, N., M. K. Poetz, and M. Schreier. 2014. Integrating Problem Solvers from Analogous Markets in New Product Ideation. *Management Science* 60 (4): 1063-1081.

Franzen, O. 1995. Trendforschung - qualitative und quantitative Ansätze. *Werbeforschung und Praxis* 40 (2): 51-37.

Freel, M. S. 2005. Patterns of Innovation and Skills in Small Firms. *Technovation* 25 (2): 123-134.

Freeman, L. C. 1979. Centrality in Social Networks Conceptual Clarification. *Social Networks* 1 (3): 215-239.

Freyssenet, M. 2011. Three Possible Scenarios for Cleaner Automobiles. *International Journal of Automotive Technology and Management* 11 (4): 300-311.

Fuchs, B., S. Arvanitis, and M. Woerter. 2011. External End Users and Innovation Performance. In *Proceedings of the DRUID Summer Conference 2011 on Innovation, Strategy, and Structure*. Copenhagen, Denmark, June 15-17, 2011.

Füller, J., K. Matzler, K. Hutter, and J. Hautz. 2012. Consumers' Creative Talent: Which Characteristics Qualify Consumers for Open Innovation Projects? An Exploration of Asymmetrical Effects. *Creativity and Innovation Management* 21 (3): 247-262.

Füller, J., R. Faullant, and K. Matzler. 2010. Triggers for Virtual Customer Integration in the Development of Medical Equipment - From a Manufacturer and a User's Perspective. *Industrial Marketing Management* 39 (8): 1376-1383.

Gambardella, A. and M. S. Giarratana. 2007. General Technologies, Product-Market Fragmentation and the Market for Technology: Evidence from the Software Security Industry. In *DRUID-DIME Academy Winter 2008 PhD conference: Conference on economics and management of innovation and organizational change*. Aalborg, Denmark, January 17-19, 2008.

Gängl-Ehrenwerth, C., R. Faullant, and E. J. Schwarz. 2013. Kundenintegration in den Neuproduktentwicklungsprozess. In *Kreativität, Innovation, Entrepreneurship*, ed. D. E. Krause, 371-384. Wiesbaden: Gabler.

Gassmann, O. and B. Gaso. 2004. Insourcing Creativity with Listening Posts in Decentralized Firms. *Creativity and Innovation Management* 13 (1): 3-14.

Gassmann, O., E. Enkel, and H. Chesbrough. 2010. The Future of Open Innovation. *R&D Management* 40 (3): 213-221.

George, G. 2005: Slack Resources and the Performance of Privately Held Firms. *Academy of Management Journal* 48 (4): 661-676.

GfK. 2013. Geschäftsbericht 2013. Nuremberg: Gesellschaft für Konsumforschung.

Ghazarian, S. and M. Nematbakhsh. 2014. Enhancing Memory-Based Collaborative Filtering for Group Recommender Systems. *Expert Systems with Applications*. Available from http://dx.doi.org/10.1016/j.eswa.2014.11.042.

Ghosh, A. and G. McGraw. 2012. Lost Decade or Golden Era: Computer Security since 9/11. *IEEE Security & Privacy* 10 (1): 6-10.

Gilley, J. W., M. L. Morris, A. M. Waite, T. Coates, and A. Veliquette. 2010. Integrated Theoretical Model for Building Effective Teams. *Advances in Developing Human Resources* 12(1): 7-28.

Goldberg, D., D. Nichols, B. M. Oki, and D. Terry. 1992. Using Collaborative Filtering to Weave an Information Tapestry. *Communications of the ACM* 35 (12): 61-70.

Goldberg, K., T. Roeder, D. Gupta, and C. Perkins. 2001. Eigentaste: A Constant Time Collaborative Filtering Algorithm. *Information Retrieval* 4 (2): 133-151.

Göthlich, S. E. 2009. Zum Umgang mit fehlenden Daten in großzahligen empirischen Erhebungen. In *Methodik der empirischen Forschung*, ed. S. Albers, D. Klapper, U. Konradt, A. Walter, and J. Wolf, 119-135. Berlin: Springer-Verlag.

Gottesdiener, E. 2003. Requirements by Collaboration: Getting It Right the First Time. *IEEE Software* 20 (2): 52-55.

Graner, M. and M. Mißler-Behr. 2013. Key determinants of the successful adoption of new product development methods. *European Journal of Innovation Management* 16 (3): 301-316.

Green, P. E. 1984. Hybrid Models for Conjoint Analysis: An Expository Review. *Journal of Marketing Research* 21 (2): 155-169.

Green, P. E. and A. M. Krieger. 1996. Individualized Hybrid Models for Conjoint Analysis. *Management Science* 42 (6): 850-867.

Green, P. E. and V. R. Rao. 1971. Conjoint Measurement for Quantifying Judgmental Data. *Journal of Marketing Research* 8 (3): 355-363.

Green, P. E. and V. Srinivasan. 1978. Conjoint Analysis in Consumer Research: Issues and Outlook. *Journal of Consumer Research* 5 (2): 103-123.

Green, P. E. and V. Srinivasan. 1990. Conjoint Analysis in Marketing: New Developments with Implications for Research and Practice. *Journal of Marketing* 54 (4): 3-19.

Green, P. E. and Y. J. Wind. 1975. New Way to Measure Consumer' Judgements. *Harvard Business Review* 53 (4): 107-115.

Green, P. E., A. M. Krieger, and M. K. Agarwal. 1991. Adaptive Conjoint Analysis: Some Caveats and Suggestions. *Journal of Marketing Research* 28 (5): 215-222.

Green, P. E., A. M. Krieger, and M. K. Agarwal. 1993. A Cross Validation Test of Four Models for Quantifying Multiattribute Preferences. *Marketing Letters* 4 (4): 369-380.

Green, P. E., A. M. Krieger, and Y. J. Wind. 2001. Thirty Years of Conjoint Analysis: Reflections and Prospects. *Interfaces* 31 (3): 56-73.

Griffin, A. and J. R. Hauser. 1993. The Voice of the Customer. *Marketing Science* 12 (1): 1-27.

Gruner, K. E. and C. Homburg. 2000. Does Customer Interaction Enhance New Product Success? *Journal of Business Research* 49 (1): 1-14.

Hall, H. and D. Graham. 2004. Creation and Recreation: Motivating Collaboration to Generate Knowledge Capital in Online Communities. *International Journal of Information Management* 24 (3): 235-246.

Harhoff, D., J. Henkel, and E. von Hippel. 2003. Profiting from Voluntary Information Spillovers: How Users Benefit by Freely Revealing Their Innovations. *Research Policy* 32 (10): 1753-1769.

Hartley, R. F. 1992. *Marketing Mistakes*. 5th Edition. New York: Wiley.

Hartmann, A. and H. Sattler. 2002. Commercial Use of Conjoint Analysis in Germany, Austria, and Switzerland. *Working Paper*. Universität Hamburg.

Hartmann, A. and H. Sattler. 2004. Wie robust sind Methoden zur Präferenzmessung? *Zeitschrift für betriebswirtschaftliche Forschung* 56 (1): 3-22.

Hauser, J., G. J. Tellis, and A. Griffin. 2006. Research on Innovation: A Review and Agenda for Marketing Science. *Marketing Science* 25 (6): 687-717.

Heeler, R. M., C. Okechuku, and S. Reid. 1979. Attribute Importance: Contrasting Measurements. *Journal of Marketing Research* 16 (1): 60-63.

Heiskanen, E., K. Hyvönen, M. Niva, M. Pantzar, P. Timonen, and J. Varjonen. 2007. User Involvement in Radical Innovation: Are Consumers Conservative? *European Journal of Innovation Management* 10 (4): 489-509.

Heiskanen, E., R. Lovio, and M. Jalas. 2011. Path Creation for Sustainable Consumption: Promoting Alternative Heating Systems in Finland. *Journal of Cleaner Production 19 (16)*: 1892-1900.

Helm, R., M. Steiner, A. Scholl, and L. Manthey. 2008. A Comparative Empirical Study on Common Methods for Measuring Preferences. *International Journal of Management and Decision Making* 9 (3): 242-265.

Henard, D. H. and D. M. Szymanski. 2001. Why Some New Products Are More Successful than Others. *Journal of Marketing Research* 38 (3): 362-375.

Henkel, J. and S. Jung. 2009. The Technology-Push Lead User Concept: A New Tool for Application Identification. *Working Paper*. Technische Universität München.

Henkel, J. and S. Jung. 2010. Identifying Technology Applications Using an Adaptation of the Lead User Method. *Working Paper*. Technische Universität München.

Henkel, J. and E. von Hippel. 2005. Welfare Implications of User Innovation. *Journal of Technology Transfer* 30 (1/2): 73-87.

Hennessey, B. A. and T. M. Amabile. 2010. Creativity. *Annual Review of Psychology* 61: 569-598.

Hennig-Thurau, T., A. Marchand, and P. Marx. 2012. Can Automated Group Recommender Systems Help Consumers Make Better Choices? *Journal of Marketing* 76 (6): 89-109.

Hensel-Börner, S. 2000. *Validität computergestützter hybrider Conjoint-Analysen.* Wiesbaden: Gabler.

Hensel-Börner, S. and H. Sattler. 1999. Validity of the Customized Computerized Conjoint Analysis (CCC). In *Proceedings of the 28th Annual Conference of the European Marketing Academy.* Berlin, Germany, 11-14 May, 1999.

Hering, S., T. Redlich, J. P. Wulfsberg, and F.-L. Bruhns. 2011. Open Innovation im Automobilbau. *Zeitschrift für Wissenschaft und Forschung* 106 (9): 647-652.

Herlocker, J. L., J. A. Konstan, L. G. Terveen, and J. T. Riedl. 2004. Evaluating Collaborative Filtering Recommender Systems. *ACM Transactions on Information Systems* 22 (1): 5-53.

Herstatt, C. 2003. Onlinegestützte Suche nach innovativen Anwendern in direkten und analogen Anwendermärkten. *Working Paper* 21. Technische Universität Hamburg-Harburg.

Herstatt, C. 2014. Implementierung innovativer Analogien in der Medizintechnik: Drei Fallstudien. In *Innovationen durch Wissenstransfer,* ed. C. Herstatt, K. Kalogerakis, and M. Schulthess, 125-136. Wiesbaden: Gabler.

Herstatt, C. and E. von Hippel. 1992. From Experience: Developing New Product Concepts via the Lead User Method: A Case Study in a "Low Tech" Field. *Journal of Product Innovation Management* 9 (3): 213-221.

Herstatt, C., C. Lüthje, and C. Lettl. 2001. Innovationsfelder mit Lead Usern erschließen. *Working Paper.* Technische Universität Hamburg-Harburg.

Herstatt, C., C. Lüthje, and C. Lettl. 2002. Wie fortschrittliche Kunden zu Innovationen stimulieren. *Harvard Business Manager* 24 (1): 60-68.

Herzog, U. 1991. *Fahrradpatente. Erfindungen aus zwei Jahrhunderten,* 2nd ed. Kiel: Moby Dick.

Hienerth, C. and C. Lettl. 2011. Exploring How Peer Communities Enable Lead User Innovations to Become Standard Equipment in the Industry: Community Pull Effects. *Journal of Product Innovation Management* 28 (S1): 175-195.

Hienerth, C., and F. Riar. 2014. Using Crowds for Evaluation Tasks: Validity by Numbers vs. Validity by Expertise Validity by Expertise. *Working Paper*. WHU-Otto Beisheim School of Management.

Hienerth, C., C. Lettl, and P. Keinz. 2013. Synergies among Producer Firms, Lead Users, and User Communities: The Case of the LEGO Producer - User Ecosystem. *Journal of Product Innovation Management* 31 (4): 848-866.

Hienerth, C., M. K. Poetz, and E. von Hippel. 2007. Exploring Key Characteristics of Lead User Workshop Participants: Who Contributes Best to the Generation of Truly Novel Solutions? In *Proceedings of the DRUID Summer Conference 2007 on Appropriability, Proximity, Routines and Innovation*. Copenhagen, Denmark, June 18-20, 2007.

Hienerth, C., E. von Hippel, and M. B. Jensen. 2014. User Community vs. Producer Innovation Development Efficiency: A First Empirical Study. *Research Policy* 43 (1): 190-20.

Hinsch, M. E., C. Stockstrom, and C. Lüthje. 2014. User Innovation in Techniques: A Case Study Analysis in the Field of Medical Devices. *Creativity and Innovation Management* 23 (4): 484-494.

Hoch, S. J. 1988. Who Do We Kow: Predicting the Interests and Opinions of the American Consumer. *Journal of Consumer Research* 15 (3): 315-324.

Horn, C. and A. Brem. 2013. Strategic Directions on Innovation Management – A Conceptual Framework. *Management Research Review* 36 (10): 939-954.

Horton, N. J. and S. R. Lipsitz. 2001. Multiple Imputation in Practice: Comparison of Software Packages for Regression Models with Missing Variables. *The American Statistician* 55 (3): 244-254.

Hoyer, W. D., R. Chandy, M. Dorotic, M. Krafft, and S. S. Singh. 2010. Consumer Cocreation in New Product Development. *Journal of Service Research* 13 (3): 283-296.

Hribernik, K. A., Z. Ghrairi, C. Hans, and K.-D. Thoben. 2011. Co-creating the Internet of Things - First Experiences in the Participatory Design of Intelligent Products with Arduino. In *17th International Conference on Concurrent Enterprising (ICE)*. Aachen, Germany, 20-22 June, 2011.

Huang, Z., H. Chen, and D. Zeng. 2004. Applying Associative Retrieval Techniques to Alleviate the Sparsity Problem in Collaborative Filtering. *ACM Transactions on Information Systems* 22 (1): 116-142.

Huber, J., D. R. Wittink, J. A. Fiedler, and R. Miller. 1993. The Effectiveness of Alternative Preference Elicitation Procedures in Predicting Choice. *Journal of Marketing Research* 30 (1): 105-114.

Huizingh, E. K. 2011. Open Innovation: State of the Art and Future Perspectives. *Technovation* 31 (1): 2-9.

Ili, S., A. Albers and S. Miller. 2010. Open Innovation in the Automotive Industry. *R&D Management* 40 (3): 246-255.

Jain, A. K., V. Mahajan, and N. K. Malhotra. 1979. Multiattribute Preference Models for Consumer Research: A Synthesis. *Advances in Consumer Research* 6 (1): 248-252.

Jaworski, B. J. and A. K. Kohli. 1993. Market Orientation: Antecedents and Consequences. *Journal of Marketing* 57 (3): 53-70.

Jensen, M. B., C. Hienerth, and C. Lettl. 2014. Forecasting the Commercial Attractiveness of User-Generated Designs Using Online Data: An Empirical Study within the LEGO User Community. *Journal of Product Innovation Management* 31 (S1): 75-93.

Jeppesen, L. B. 2005. User Toolkits for Innovation: Consumers Support Each Other. *Journal of Product Innovation Management* 22 (4): 347-362.

Jeppesen, L. B. and L. Frederiksen. 2006. Why Do Users Contribute to Firm-Hosted User Communities? The Case of Computer-Controlled Music Instruments. *Organization Science* 17 (1): 45-63.

Jeppesen, L. B. and K. R. Lakhani. 2010. Marginality and Problem-Solving Effectiveness in Broadcast Search. *Organization Science* 21 (5): 1016-1033.

Jeppesen, L. B. and K. Laursen. 2009. The Role of Lead Users in Knowledge Sharing. *Research Policy* 38 (10): 1582-1589.

Johnson, R. D. and I. P. Levin. 1985. More than meets the eye: The effect of missing information on purchase evaluations. *Journal of Consumer Research* 12 (2): 169–177.

Johnson, R. M. 1974. Trade-Off Analysis of Consumer Values. *Journal of Marketing Research* 11 (2): 121-127.

Johnson, R. M. 1987. Adaptive Conjoint Analysis. In *Conference Proceedings of Sawtooth Software Conference on Perceptual Mapping, Conjoint Analysis, and Computer Interviewing*. Sun Valley, USA, March, 1987.

Johnson, R. M. 1991. Comment on "Adaptive Conjoint Analysis: Some Caveats and Suggestions". *Journal of Marketing Research* 28 (2): 223-225.

Johnson, R. M. and B. K. Orme. 2007. A New Approach to Adaptive CBC. *Working Paper*. Sawtooth Software Inc.

Kalogerakis, K., C. Lüthje, and C. Herstatt. 2010. Developing Innovations Based on Analogies: Experience from Design and Engineering Consultants. *Journal of Product Innovation Management* 27 (3): 418-436.

Kash, D. E. and R. W. Rycoft 2000. Patterns of innovating complex technologies: a framework for adaptive network strategies. *Research Policy* 29 (7): 819-831.

Kathoefer, D. G., and J. Leker. 2012. Knowledge Transfer in Academia: An Exploratory Study on the Not-Invented-Here Syndrome. *Journal of Technology Transfer* 37 (5): 658-675.

Katz, R. and T. J. Allen. 1982. Investigating the Not Invented Here (NIH) Syndrome: A Look at the Performance, Tenure, and Communication Patterns of 50 R&D Project Groups. *R&D Management* 12 (1): 7-20.

Keinz, P. and R. Prügl. 2010. A User Community-Based Approach to Leveraging Technological Competences: An Exploratory Case Study of a Technology Start-Up from MIT. *Creativity and Innovation Management* 19 (3): 269-289.

Khurana, A. and S. R. Rosenthal. 1997. Integrating the Fuzzy Front End of New Product Development. *Sloan Management Review* 38 (2): 103-120.

Kietzmann, J. H., and I. Angell. 2014. Generation-C: Creative Consumers in a World of Intellectual Property Rights. *International Journal of Technology Marketing* 9 (1): 86-98.

Kim, J. and D. Wilemon. 2002. Focusing the Fuzzy Front–End in New Product Development. *R&D Management* 32 (4): 269-279.

Kim, J.-O. and J. Curry. 1977. The Treatment of Missing Data in Multivariate Analysis. *Sociological Methods & Research* 6 (2): 215-240.

Klebe-Klingemann, R., and C. Schneider. 2010. Fernwirkcontroller verbindet klassische Automation mit Fernwirken. *Netzpraxis* 49 (5): 20-23.

Kleinschmidt, E. J. and R. G. Cooper. 1991. The Impact of Product Innovativeness on Performance. *Journal of Product Innovation Management* 8 (4): 240-251.

König, H. 2005. Peer-to-Peer Intrusion Detection Systeme für den Schutz sensibler IT-Infrastrukturen. In *Beiträge der 35. Jahrestagung der Gesellschaft für Informatik e.V. (GI)*. Bonn, Germany, September 19-22, 2005.

Kornish, L. J. and K. T. Ulrich. 2011. Opportunity Spaces in Innovation: Empirical Analysis of Large Samples of Ideas. *Management Science* 57 (1): 107-12.

Kornish, L. J. and K. T. Ulrich. 2014. The Importance of the Raw Idea in Innovation: Testing the Sow's Ear Hypothesis. *Journal of Marketing Research* 51 (1): 14-26.

Kozinets, R. V. 2002. The Field behind the Screen: Using Netnography for Marketing Research in Online Communities. *Journal of Marketing Research* 39 (1): 61-72.

Kratzer, J. and C. Lettl. 2009. Distinctive Roles of Lead Users and Opinion Leaders in the Social Networks of Schoolchildren. *Journal of Consumer Research* 36 (4): 646-659.

Krimmling, J. and P. Langendörfer. 2014. Intrusion Detection Systems for (Wireless) Automation Systems. In *The State of the Art in Intrusion Prevention and Detection*, ed. A.-S. K. Pathan, 431-448. Boca Raton: CRC Press.

Krimmling, J. and A. Sänn. 2015. Der Gefahr trotzen - Erweiterte Sicherheit für industrielle Anlagen. *chemie&more* 1/2015: 18-21.

Krishnan, V., P. K. Narayanashetty, M. Nathan, R. T. Davies, and J. A. Konstan. 2008. Who Predicts Better? - Results from an Online Study Comparing Humans and an Online Recommender System. In *Proceedings of the 2nd ACM Conference on Recommender Systems*. Lousanne, Switzerland, October 23-25, 2008.

Kroeber-Riel, W., P. Weinberg, and A. Gröppel-Klein. 2009. *Konsumentenverhalten* 9th ed. München: Verlag Franz Vahlen.

Krzywicki, A., W. Wobcke, Y. S. Kim, X. Cai, M. Bain, A. Mahidadia, and P. Compton, P. 2015. Collaborative Filtering for People-to-People Recommendation in Online Dating: Data Analysis and User Trial. *International Journal of Human-Computer Studies* 76 (1): 50-66.

Kutschke, A. 2014. *Erfolgsfaktoren innovativer Energietechnologien – Eine produkt- kooperations- und standortbezogene Betrachtung.* Hamburg: Verlag Dr. Kovač.

Lakhani, K. R. 2006. Broadcast Search in Problem Solving: Attracting Solutions from the Periphery. In *Proceedings of Technology Management for the Global Future PICMET 2006*, Istanbul, Turkey, July 8-13, 2006.

Lakhani, K. R. and R. G. Wolf. 2005. Why Hackers Do What They Do: Understanding Motivation and Effort in Free/Open Source Software Projects. In *Perspectives on Free and Open Source Software*, ed. J. Feller, B. Fitzgerald, S. Hissam, and K. R. Lakhani, 3-22. Cambridge: MIT Press.

Lang, A. 2006. Der Kunde als „Entwicklungs"-Helfer. *Absatzwirtschaft - Zeitschrift für Marketing* 49 (2): 36-38.

Lang, A. and S. Reich. 2008. „Outside in" - Die erfolgreiche Integration von Endkunden in den Innovationsprozess. In *Die neue Macht des Marketing*, ed. R. T. Kreutzer, and W. Merkle, 131-147. Wiesbaden: Gabler.

Langner, R. 2011. Stuxnet: Dissecting a Cyberwarfare Weapon. *IEEE Computer and Reliability Societies* 9 (3): 49-51.

Larkin, R. D., J. Lopez, J. W. Butts, and M. R. Grimaila. 2014. Evaluation of Security Solutions in the SCADA Environment. *The DATA BASE for Advances in Information Systems* 45 (1): 38-53.

Lasagni, A. 2012. How Can External Relationships Enhance Innovation in SMEs? New Evidence for Europe. *Journal of Small Business Management* 50 (2): 310-339.

Lazzarotti, V., R. Manzini, L. Pellegrini, and E. Pizzurno. 2013. Open Innovation in the Automotive Industry: Why and How? Evidence from a Multiple Case

Study. *International Journal of Technology Intelligence and Planning* 9 (1): 37-56.

Lehnen, J., D. Ehls, and C. Herstatt. 2014. Implementation of Lead Users into Management Practice – A Literature Review of Publications in Business Press. *Working Paper.* Technische Universität Hamburg-Harburg.

Leigh, T. W., D. B. MacKay, and J. O. Summers. 1984. Reliability and Validity of Conjoint Analysis and Self-Explicated Weights: A Comparison. *Journal of Marketing Research* 21 (4): 456-462.

Lettl, C., C. Herstatt, and H. G. Gemünden. 2006a. Learning from Users for Radical Innovation. *International Journal of Technology Management* 33 (1): 25-45.

Lettl, C., C. Herstatt, and H. G. Gemünden. 2006b. Users' Contributions to Radical Innovation: Evidence from Four Cases in the Field of Medical Equipment Technology. *R&D Management* 36 (3): 251-272.

Lettl, C., K. Rost, and I. von Wartburg. 2009. Why are some Independent Inventors 'Heroes' and Others 'Hobbyists'? The Moderating Role of Technological Diversity and Specialization. *Research Policy* 38 (2): 243-254.

Lichtenthaler, E. 2004. Technology Intelligence Processes in Leading Europe and North American Multinationals. *R&D Management* 34 (2): 121-135.

Lichtenthaler, E. 2005. The Choice of Technology Intelligence Methods in Multinationals: Towards a Contingency Approach. *International Journal of Technology Management* 32 (3/4): 388-405.

Lichtenthaler, U. and H. Ernst. 2006. Attitudes to Externally Organising Knowledge Management Tasks: A Review, Reconsideration and Extension of the NIH Syndrome. *R&D Management* 36 (4): 367-386.

Lilien, G. L., P. D. Morrison, K. Searls, M. Sonnack, E. von Hippel. 2002. Performance Assessment of the Lead User Idea-Generation Process for New Product Development. *Management Science* 48 (8): 1042-1059.

Lin, B.-W. 2003. Technology Transfer as Technological Learning: A Source of Competitive Advantage for Firms with Limited R&D Resources. *R&D Management* 33 (3): 327-341.

Linden, G., B. Smith, and J. York. 2003. Amazon.com Recommendations: Item-to-Item Collaborative Filtering. *IEEE Internet Computing* 7 (1): 76-80.

Little, R. J. A. 1982. Models for Nonresponse in Sample Surveys. *Journal of the American Statistical Association* 77 (378): 237-250.

Little, R. J. A. 1988. Robust Estimation of the Mean and Covariance Matrix from Data with Missing Values. *Journal of the Royal Statistical Society. Series C (Applied Statistics)* 37 (1): 23-38.

Little, R. J. A. and D. B. Rubin. 2014. *Statistical Analysis with Missing Data*. New York: Wiley.

Louviere, J. J. and G. Woodworth. 1983. Design and Analysis of Simulated Consumer Choice or Allocation Experiments: An Approach Based on Aggregate Data. *Journal of Marketing Research* 20 (4): 350-367.

Luce, R. D. and J. W. Tukey. 1964. Simultaneous Conjoint Measurement: A New Type of Fundamental Measurement. *Journal of Mathematical Psychology* 1 (1): 1-27.

Luchins, A. S. 1942. Mechanization in Problem Solving: The Effect of "Einstellung". *Psychological Monographs* 54 (6): i-95.

Lüthje, C. 2000. *Kundenorientierung im Innovationsprozess: Eine Untersuchung der Kunden-Hersteller-Interaktion in Konsumgütermärkten*. Wiesbaden: Deutscher Universitäts-Verlag.

Lüthje, C. 2003. Customers as Co-Inventors: An Empirical Analysis of the Antecedents of Customer-Driven Innovations in the Field of Medical Equipment. In *Proceedings of the 32nd Annual Conference of the European Marketing Academy (EMAC)*, UK, Glasgow, May 20-23, 2003.

Lüthje, C. 2004. Characteristics of Innovating Users in a Consumer Goods Field: An Empirical Study of Sport-Related Product Consumers. *Technovation* 24 (9): 683-695.

Lüthje, C. and C. Herstatt. 2004. The Lead User Method: An Outline of Empirical Findings and Issues for Future Research. *R&D Management* 34 (5): 553-568.

Lüthje, C., C. Herstatt, and E. von Hippel. 2005. User-Innovators and "Local" Information: The Case of Mountain Biking. *Research Policy* 34 (6): 951-965.

Lüthje, C., C. Lettl, and C. Herstatt. 2003. Knowledge Distribution among Market Experts: A Closer Look into the Efficiency of Information Gathering for Innovation Projects. *Working Paper.*

Magnusson, P. R. 2009. Exploring the Contributions of Involving Ordinary Users in Ideation of Technology-Based Services. *Journal of Product Innovation Management* 26 (5): 578-593.

Mahr, D. and A. Lievens. 2012. Virtual Lead User Communities: Drivers of Knowledge Creation for Innovation. *Research Policy* 41 (1): 167-177.

Manfreda, K. L., M. Bosnjak, J. Berzelak, I. Haas, and V. Vehovar. 2008. Web Surveys versus other Survey Modes: A Meta-Analysis Comparing Response Rates. *International Journal of Market Research* 50 (1): 79-104.

Mansfield, E. 1968. *Industrial Research and Technological Innovation: An Econometric Analysis.* New York: Norton.

Marchi, G., C. Giachetti, and P. de Gennaro. 2011. Extending Lead-User Theory to Online Brand Communities: The Case of the Community Ducati. *Technovation* 31 (8): 350-361.

Markham, S. K., and H. Lee. 2013. Product Development and Management Association's 2012 Comparative Performance Assessment Study. *Journal of Product Innovation Management* 30 (3): 408-429.

Martinez, M. G. and B. Walton. 2014. The Wisdom of Crowds: The Potential of Online Communities as a Tool for Data Analysis. *Technovation* 34 (4): 203-214.

Maxton, G. P. and J. Wormald. 2004. *Time for a Model Change: Re-engineering the Global Automotive Industry.* Cambridge, UK: Cambridge University Press.

Mazis, M. B., O. T. Ahtola, and R. E. Klippel. 1975. A Comparison of Four Multi-Attribute Models in the Prediction of Consumer Attitudes. *Journal of Consumer Research* 2 (1): 38-52.

McFadden, D. 1986. The Choice Theory Approach to Market Research. *Marketing Science* 5 (4): 275-297.

McGee, J., H. Thomas, and D. Wilson. 2005. *Strategy: Analysis and Practice.* New York: McGraw-Hill.

Meehan, S. and P. Baschera. 2002. Lessons from Hilti: How Customer and Employee Contact Improves Strategy Implementation. *Business Strategy Review* 13 (2): 31-39.

Meinberg, U. 1989. *Steuerung von fahrerlosen Transportsystemen: Regelwerk zum rechnergestützten Entwurf.* Köln: Verlag TÜV Rheinland.

Meißner, H.-R. 2012. *Preisdruck auf die Automobil-Zulieferindustrie.* Berlin: Forschungsgemeinschaft für Außenwirtschaft, Struktur- und Technologiepolitik e.V.

Meißner, M., R. Decker, and N. Adam. 2011. Ein empirischer Validitätsvergleich zwischen Adaptive Self-Explicated Approach (ASE), Pairwise Comparison-based Preference Measurement (PCPM) und Adaptive Conjoint Analysis (ACA). *Zeitschrift für Betriebswirtschaft* 81 (4): 423-446.

MIFA. 2013. *2012 Geschäftsbericht.* Sangerhausen: MIFA Mitteldeutsche Fahrradwerke AG.

Milgram, S. 1967. The Small-World Problem. *Psychology Today* 2 (1): 60-67.

Molitor, A. 2007. Der Dicke - Es sollte das beste Auto der Welt sein. Es wurde das unbeliebteste seiner Klasse. *brand eins* 07(07): 92-97.

Morrison, P. D. 1995. Testing a Framework for the Adoption of Technological Innovations by Organizations and the Role of Leading Edge Users. *Working Paper.* University of Sydney.

Morrison, P. D., J. H. Roberts, and D. F. Midgley. 2004. The Nature of Lead Users and Measurement of Leading Edge Status. *Research Policy* 33 (2): 351-362.

Morrison, P. D., J. H. Roberts, and E. von Hippel. 2000. Determinants of User Innovation and Innovation Sharing in a Local Market. *Management Science* 46 (12): 1513-1527.

Morrison, P. D., J. H. Roberts, D. F. Midgley. 1999. Towards a Finer Understanding of Lead Users. *Working Paper.* The Pennsylvania State University.

Moser, K. and F. T. Piller. 2006. The International Mass Customisation Case Collection: An Opportunity for Learning from previous Experiences. *International Journal of Mass Customisation* 1 (4): 403-409.

Müller, A. 2008. *Strategic Foresight - Prozesse strategischer Trend- und Zukunftsforschung in Unternehmen*. PhD Thesis. Universität St. Gallen.

Nambisan, S. and R. A. Baron. 2010. Different Roles, Different Strokes: Organizing Virtual Customer Environments to Promote Two Types of Customer Contributions. *Organization Science* 21 (2): 554-572.

Narver, J. C. and S. F. Slater. 1990. The Effect of a Market Orientation on Business Profitability. *Journal of Marketing* 54 (4): 20-35.

Narver, J. C., S. F. Slater, and D. L. MacLachlan. 2004. Responsive and Proactive Market Orientation and New-Product Success. *Journal of Product Innovation Management* 21 (5): 334-347.

Natalicchio, A., A. M. Petruzzelli, and A. Garavelli. 2014. A Literature Review on Markets for Ideas: Emerging Characteristics and Unanswered Questions. *Technovation* 34 (2): 65-76.

Nathanson, T., E. Bitton, and K. Goldberg. 2007. Eigentaste 5.0: Constant-Time Adaptability in a Recommender System Using Item Clustering. In *Proceedings of the 1st ACM Conference on Recommender systems*. Minneapolis, USA, October 19-20, 2007.

Neibecker, B. and T. Kohler. 2009. Produktdesign auf Basis von Conjointdaten. In *Conjointanalyse*, ed. D. Baier, and M. Brusch, 215-232. Berlin: Springer-Verlag.

Netzer, O. and V. Srinivasan. 2011. Adaptive Self-Explication of Multiattribute Preferences. *Journal of Marketing Research* 48 (1): 140-156.

Netzer, O., O. Toubia, E. T. Bradlow, E. Dahan, T. Evgeniou, F. M. Feinberg, E. M. Feit, S. K. Hui, J. Johnson, J. C. Liechty, J. B. Orlin, and V. R. Rao. 2008. Beyond Conjoint Analysis: Advances in Preference Measurement. *Marketing Letters* 19 (3/4): 337-354.

Nishikawa, H., M. Schreier, and S. Ogawa. 2013. User-Generated Versus Designer-Generated Products: A Performance Assessment at Muji. *International Journal of Research in Marketing* 30 (2): 160-167.

Olausson, D. and C. Berggren. 2010. Managing Uncertain, Complex Product Development in High-Tech Firms: In Search of Controlled Flexibility. *R&D Management* 40 (4): 383-399.

Oliveira, P. and E. von Hippel. 2011. Users as Service Innovators: The Case of Banking Services. *Research Policy* 40 (6): 806-818.

Olson, E. L. and G. Bakke. 2001. Implementing the Lead User Method in a High Technology Firm: A Longitudinal Study of Intentions versus Actions. *Journal of Product Innovation Management* 18 (6): 388-395.

Olson, E. L. and G. Bakke. 2004. Creating Breakthrough Innovations by Implementing the Lead User Methodology. *Telektronikk* 13 (2): 126-132.

Oppewal, H. and M. Klabbers. 2003. Compromising between Information Completeness and Task Simplicity: A Comparison of Self-Explicated, Hierarchical Information Integration, and Full-profile Conjoint Methods. *Advances in Consumer Research* 30 (1): 298-304.

Owens, J. D. 2007. Why Do Some UK SMEs Still Find the Implementation of a New Product Development Process Problematical?: An Exploratory Investigation. *Management Decision* 45 (2): 235-251.

Ozer, M. 2009. The Roles of Product Lead-Users and Product Experts in New Product Evaluation. *Research Policy* 38 (8): 1340-1349.

Pajo, S., P.-A. Verhaegen, D. Vandevenne, and J. R. Duflou. 2013. Analysis of Automatic Online Lead User Identification. In *Smart Product Engineering*, ed. M. Abramovici, and R. Stark, 505-514. Berlin: Springer-Verlag.

Paradiso, J. A., J. Heidemann, and T. G. Zimmerman. 2008. Hacking is Pervasive. *IEEE Pervasive Computing* 7 (3): 13-15.

Parida, V., M. Westerberg, and J. Frishammar. 2012. Inbound Open Innovation Activities in High-Tech SMEs: The Impact on Innovation Performance. *Journal of Small Business Management* 50 (2): 283-309.

Park, Y.-H., M. Ding, and V. R. Rao. 2008. Eliciting Preference for Complex Products: A Web-Based Upgrading Method. *Journal of Marketing Research* 45 (5): 562-574.

Pathan, A.-S. K. 2014. *The State of the Art in Intrusion Prevention and Detection.* Boca Raton: CRC Press.

Pauwels, K., J. Silva-Risso, S. Srinivasan, and D. M. Hanssens. 2004. New Products, Sales Promotions, and Firm Value: The Case of the Automobile Industry. *Journal of Marketing* 68 (4): 142-156.

Piller, F. T. and D. Walcher. 2006. Toolkits for Idea Competitions: A Novel Method to Integrate Users in New Product Development. *R&D Management* 36 (3): 307-318.

Poetz, M. K. and R. Prügl. 2010. Crossing Domain-Specific Boundaries in Search of Innovation: Exploring the Potential of Pyramiding. *Journal of Product Innovation Management* 27 (6): 897-914.

Poetz, M. K. and M. Schreier. 2012. The Value of Crowdsourcing: Can Users Really Compete with Professionals in Generating New Product Ideas? *Journal of Product Innovation Management* 29 (2): 245-256.

Poetz, M. K., C. Steger, I. Mayer, and J. Schrampf. 2005. Evaluierung von Case Studies zur Lead User Methode. *Working Paper.* ECONSULT Betriebsberatungsges.m.b.H.

Pollock, T. G. and J. E. Bono. 2013. Being Scheherazade: The Importance of Storytelling in Academic Writing. *Academy of Management Journal* 56 (3): 629-634.

Portilla, J., J. A. Otero, E. de la Torre, T. Riesgo, O. Stecklina, S. Peter, and P. Langendörfer. 2010. Adaptable Security in Wireless Sensor Networks by Using Reconfigurable ECC Hardware Coprocessors. *International Journal of Distributed Sensor Networks.* Available from http://dx.doi.org/10.1155/2010/740823.

Pullman, M. E., K. J. Dodson, and W. L. Moore. 1999. A Comparison of Conjoint Methods. When There Are Many Attributes. *Marketing Letters* 10 (2): 1-14.

Raasch, C., C. Herstatt, and P. Lock. 2008. The Dynamics of User Innovation: Drivers and Impediments of Innovation Activities. *International Journal of Innovation Management* 12 (3): 377-398.

Rammer, C., B. Aschhoff, D. Crass, T. Doherr, M. Hud, C. Köhler, B. Peters, T. Schubert, and F. Schwiebacher. 2012. *Innovationsverhalten der deutschen Wirtschaft - Indikatorenbericht zur Innovationserhebung 2011*. Mannheim: Zentrum für europäische Wirtschaftsforschung GmbH.

Rao, V. R. 2014. *Applied Conjoint Analysis*. Berlin: Springer-Verlag.

Reger, G. and C. Schultz. 2009. Lead-Using or Lead-Refusing? An Examination of Customer Integration in Mechanical Engineering Firms International. *Journal of Technology Marketing* 4 (2-3): 217-229.

Reichwald, R. and F. Piller. 2009. *Interaktive Wertschöpfung: Open Innovation, Individualisierung und neue Formen der Arbeitsteilung*. Wiesbaden: Gabler.

Rese, A. and D. Baier. 2011. Success Factors for Innovation Management in Networks of Small and Medium Enterprises. *R&D Management* 41 (2): 138-155.

Rese, A., A. Sänn, and F. Homfeldt. 2015. Customer Integration and Voice-of-Customer Methods in the German Automotive Industry. *International Journal of Automotive Technology and Management* 15 (1): 1-19.

Resnick, P., N. Iacovou, M. Suchak, P. Bergstrom, and J. Riedl. 1994. GroupLens: An Open Architecture for Collaborative Filtering of Netnews. In *Proceedings of the 5th ACM Conference on Computer Supported Cooperative Work*. Chapel Hill, USA, October 22-26, 1994.

Richter, K. and P. Hartig. 2007. Aufbau globaler Netzwerke als Erfolgsfaktor in der Automobilindustrie. In *Die Automobilindustrie auf dem Weg zur globalen Netzwerkkompetenz*, ed. F. J. G. Sanz, K. Semmler, and J. Walther, 251-264. Berlin: Springer-Verlag.

Roberts, J. H., and G. L. Urban. 1985. New Consumer Durable Brand Choice: Modeling Multiattribute Utility, Risk, and Dynamics. *Working Paper*. MIT Sloan School of Management.

Robertson, A. 1974. Innovation Management: Theory and Comparative Practice Illustrated by two Case Studies. *Management Decision* 12 (6): 330-368.

Rogers, E. M. 1962. *Diffusion of Innovations*. New York: The Free Press.

Rogers, E. M. 2003. *Diffusion of Innovations* 5th ed. New York: The Free Press.

Rogers, E. M. and F. F. Shoemaker. 1971. *Communication of innovations: A cross-cultural approach* 2nd ed. New York: The Free Press.

Rosenberg, M. J. 1956. Cognitive Structure and Attitudinal Affect. *Journal of Abnormal and Social Psychology* 53 (3): 376-82.

Rosenkopf, L. and P. Almeida. 2003. Overcoming Local Search through Alliances and Mobility. *Management Science* 49 (3): 751-766.

Rothwell, R., C. Freeman, A. Horlsey, V. T. P. Jervis, A. B. Robertson, and J. Townsend. 1974. SAPPHO Updated - Project SAPPHO Phase II. *Research Policy* 3 (3): 258-291.

Rubin, D. B. 1977. Formalizing Subjective Notions about the Effect of Nonrespondents in Sample Surveys. *Journal of the American Statistical Association* 72 (359): 538-543.

Rycroft, R. W., and D. E. Kash 1994. Complex technology and community: Implications for policy and social science. *Research Policy* 23 (6): 613-626.

Saaty, T. L. 1986. Axiomatic Foundation of the Analytic Hierarchy Process. *Management Science* 32 (7): 841-855.

Saaty, T. L. 1994a. Highlights and Critical Points in the Theory and Application of the Analytic Hierarchy Process. *European Journal of Operational Research* 74 (3): 426-447.

Saaty, T. L. 1994b. How to Make a Decision: The Analytic Hierarchy Process. *Interfaces* 24 (6): 19-43.

Sánchez-González, G., N. González-Álvarez, and M. Nieto. 2009. Sticky Information and Heterogeneous Needs as Determining Factors of R&D Cooperation with Customers. *Research Policy* 38 (10): 1590-1603.

Sandberg, H., S. Amin, and K. H. Johansson. 2015. Cyberphysical Security in Networked Control Systems: An Introduction to the Issue. *IEEE Control Systems Magazine* 35 (1): 20-23.

Sänn, A. 2011. Klasse statt Masse - Wie Unternehmen Lead-User in den Innovationsprozess integrieren. *Innovationsmanager* 16: 66-67.

Sänn, A. and D. Baier. 2012. Lead User Identification in Conjoint Analysis Based Product Design. In *Challenges at the Interface of Data Analysis, Computer*

Science, and Optimization, ed. W. A. Gaul, A. Geyer-Schulz, L. Schmidt-Thieme and J. Kunze, 521-528. Heidelberg: Springer.

Sänn, A. and D. Baier. 2014. Lead Users and Non-Lead Users: Reconsidering Formal Procedures for Breakthrough Preference Measurement. *Working Paper.* Brandenburgische Technische Universität Cottbus.

Sänn, A. and J. Krimmling. 2014. Neue Wege für die IT-Sicherheit. *a+s - zeitschrift für automation und security* 3 (1): 27-29.

Sänn, A. and M. Ni. 2013. Complex Product Development: Using a Combined VoC Lead User Approach for SMEs Requirements. In *Proceedings of the 15th General Online Research Conference,* Mannheim, Germany, March 4-6, 2013.

Sänn, A., M. Keil, W. Rogmans, and I. Krebs. 2014. Die primäre Verletzungsprävention beim Umgang mit Produkten: ein Paradigmenwechsel. *das Krankenhaus* 56 (2): 144-148.

Sänn, A., J. Krimmling, D. Baier, and M. Ni. 2013. Lead User Intelligence for Complex Product Development: The Case of Industrial IT-Security Solutions. *International Journal of Technology Intelligence and Planning* 9 (3): 232-249.

Sarwar, B., G. Karypis, J. Konstan, and J. Riedl. 2000. Analysis of Recommendation Algorithms for E-Commerce. In *Proceedings of the 2nd ACM Conference on Electronic Commerce.* Minneapolis, USA, October 17-20, 2000.

Sattler, H. 2006. Methoden zur Messung von Präferenzen für Innovationen. *Zeitschrift für betriebswirtschaftliche Forschung* 54 (6): 154-176.

Sattler, H. and S. Hensel-Börner. 2007. A Comparison of Conjoint Measurement with Self-Explicated Approaches. In *Conjoint Measurement*, ed. A. Gustafsson, A. Herrmann, and F. Huber, 67-76. Berlin: Springer-Verlag.

Sawhney, M., R. C. Wolcott, and I. Arroniz. 2006. The 12 Different Ways for Companies to Innovate. *MIT Sloan Management Review* 47 (3): 75-81.

Schade, W., C. Zanker, A. Kühn, S. Kinkel, A. Jäger, T. Hettesheimer, and T. Schmall. 2012. *Zukunft der Automobilindustrie (Future of the Automotive Industry).* Berlin: Büro für Technikfolgen-Abschätzung beim Deutschen Bundestag.

Schein, A. I., A. Popescul, L. H. Ungar, and D. M. Pennock. 2002. Methods and Metrics for Cold-Start Recommendations. In *Proceedings of the 25th Annual International ACM SIGIR Conference on Research and Development in Information Retrieval (SIGIR'02)*. New York, USA, August 11-15, 2002.

Schild, K., C. Herstatt, and C. Lüthje. 2004. How to Use Analogies for Breakthrough Innovations. *Working Paper*. Technische Universität Hamburg-Harburg.

Schlereth, C., C. Eckert, R. Schaaf, and B. Skiera. 2014. Measurement of Preferences with Self-Explicated Approaches: A Classification and Merge of Trade-Off- and Non-Trade-Off-Based Evaluation Types. *European Journal of Operational Research* 238 (1): 185-198.

Schmidt, S. 2009. *Die Diffusion komplexer Produkte und Systeme: Ein systemdynamischer Ansatz*. Wiesbaden: Gabler.

Schmittlein, D. C., and V. Mahajan. 1982. Maximum Likelihood Estimation for an Innovation Diffusion Model of New Product Acceptance. *Marketing Science* 1 (1), 57-78.

Schmookler, J. 1966. *Invention and Economic Growth*. Cambridge: Harvard University Press.

Schneider, J. and J. Hall. 2011. Why Most Product Launches Fail. *Harvard Business Review* 89 (4): 21-23.

Schnell, R., P. B. Hill, and E. Esser. 1999. *Methoden der empirischen Sozialforschung*. München: Oldenbourg-Verlag.

Scholz, S. W., M. Meissner, and R. Decker. 2010. Measuring Consumer Preferences for Complex Products: A Compositional Approach Based on Paired Comparisons. *Journal of Marketing Research* 47 (4): 685-698.

Schreier, M. and R. Prügl. 2008. Extending Lead-User Theory: Antecedents and Consequences of Consumers' Lead Userness. *Journal of Product Innovation Management* 25 (4): 331-346.

Schreier, M., S. Oberhauser, and R. Prügl. 2007. Lead Users and the Adoption and Diffusion of New Products: Insights from Two Extreme Sports Communities. *Marketing Letters* 18 (1/2):15-30.

Schuhmacher, M. C., and S. Kuester. 2012. Identification of Lead User Characteristics Driving the Quality of Service Innovation Ideas. *Creativity and Innovation Management* 21 (4): 427-442.

Schumpeter, J. A. 1934. *The theory of economic development: An inquiry into profits, capital, credit, interest, and the business cycle.* New Brunswick: Transaction Publishers.

Schumpeter, J. A. 1942. *Capitalism, socialism and democracy.* New York: Harper.

Schuurman, D., D. Mahr, and L. D. Marez. 2011. User Characteristics for Customer Involvement in Innovation Processes: Deconstructing the Lead User-Concept. In *Proceedings of the ISPIM 22nd conference: Sustainability in Innovation: Innovation Management Challenges.* Hamburg, Germany, June 12-15, 2011.

Schweisfurth, T. G., and C. Raasch. 2012. Lead Users as Firm Employees: How Are They Different and Why Does It Matter? *Working Paper.* Technische Universität Hamburg-Harburg.

Schweisfurth, T. G. and C. Raasch. 2015. Embedded Lead Users - The Benefits of Employing Users for Corporate Innovation. *Research Policy* 44 (1): 168-180.

Selka, S. 2013. *Validität computergestützter Verfahren der Präferenzmessung Eine meta-analytische Untersuchung und Vorstellung neuer conjointanalytischer Lösungsansätze.* Hamburg: Verlag Dr. Kovač.

Selka, S. and D. Baier. 2014. Kommerzielle Anwendung auswahlbasierter Verfahren der Conjointanalyse: Eine empirische Untersuchung zur Validitätsentwicklung. *Marketing - Zeitschrift für Forschung und Praxis* 36 (1): 54-64.

Shah, S. K. and M. Tripsas. 2007. The Accidental Entrepreneur: The Emergent and Collective Process of User Entrepreneurship. *Strategic Entrepreneurship Journal* 1 (1-2): 123-140.

Sharma, A., M. Vardhan, and D. S. Kushwaha. 2014. A Versatile Approach for the Estimation of Software Development Effort Based on SRS Document. *International Journal of Software Engineering and Knowledge Engineering* 24 (1): 1-42.

Shi, Y., M. Larson, and A. Hanjalic. 2014. Collaborative Filtering beyond the User-Item Matrix: A Survey of the State of the Art and Future Challenges. *ACM Computing Surveys* 47 (1): Article No. 3, 1-45.

Shocker, A. D. and V. Srinivasan. 1974. A Consumer-Based Methodology for the Identification of New Product Ideas. *Management Science* 20 (6): 921-937.

Shocker, A. D. and V. Srinivasan. 1979. Multiattribute Approaches for Product Concept Evaluation and Generation: A Critical Review. *Journal of Marketing Research* 16 (2): 159-180.

Silk, A. J. and G. L. Urban. 1978. Pre-Test-Market Evaluation of New Packaged Goods: A Model and Measurement Methodology. *Journal of Marketing Research* 15 (2): 171-191.

Slater, S. F. and J. C. Narver. 1998. Customer-Led and Market-Oriented: Let's Not Confuse the Two. *Strategic Management Journal* 19 (10): 1001-1006.

Soukhoroukova, A., M. Spann, and B. Skiera. 2012. Sourcing, Filtering, and Evaluating New Product Ideas: An Empirical Exploration of the Performance of Idea Markets. *Journal of Product Innovation Management* 29 (1): 100-112.

Spann, M., H. Ernst, B. Skiera, and J. H. Soll. 2009. Identification of Lead Users for Consumer Products via Virtual Stock Markets. *Journal of Product Innovation Management* 26 (3): 322-335.

Srinivasan, V. 1988. A Conjunctive-Compensatory Approach to the Self-Explication of Multiattributed Preferences. *Decision Sciences* 19 (2): 295-305.

Srinivasan, V. and C. S. Park. 1997. Surprising Robustness of the Self-Explicated Approach to Customer Preference Structure Measurement. Journal of Marketing Research 34 (2):286-291.

Stahl, E. 2007. *Dynamik in Gruppen. Handbuch der Gruppenleitung,* 2nd ed. Weinheim: Beltz.

Stecklina, O. and A. Sänn. 2012. Erweiterte Sicherheit für Kritische Infrastrukturen (ESCI). In *Innovationsforum ISI4people.* Cottbus, Germany, June 21-22, 2012.

Steenkamp, J.-B. E., and D. R. Wittink. 1994. The Metric Quality of Full-Profile Judgments and the Number-of-Attribute-Levels Effect in Conjoint Analysis. *International Journal of Research in Marketing* 11 (3): 275-286.

Stockstrom, C. S., R. C. Goduscheit, J. Jorgensen, and C. Lüthje. 2012. Identification of Individuals with Special Qualities - Assessing the Performance of Pyramiding Search. In *Proceedings of the DRUID Society Conference 2012 on Innovation and Competitiveness - Dynamics of organizations, industries, systems and regions*. Copenhagen, Denmark, June 19-21, 2012.

Stuart, T. E. and J. M. Podolny. 1996. Local Search and the Evolution of Technological Capabilities. *Strategic Management Journal* 17 (Summer): 21-38.

Stüber, E. 2011. *Personalisierung im Internethandel*. Wiesbaden: Gabler.

Stuckmann, P. and R. Zimmermann. 2009. European Research on Future Internet Design. *IEEE Wireless Communications* 16 (5): 14-22.

Su, X. and T. M. Khoshgoftaar. 2009. A Survey of Collaborative Filtering Techniques. *Advances in Artificial Intelligence*. Available from http://dx.doi.org/ 10.1155/2009/421425.

Teichert, T. and E. Shehu. 2010. Investigating Research Streams of Conjoint Analysis: A Bibliometric Study. *Business Research* 3 (1): 49-68.

Thomke, S. and E. von Hippel 2002. Customers as Innovators: A New Way to Create Value. *Harvard Business Review* 80 (4): 74-81.

Tidd, J. 1997. Complexity, Networks & Learning: Integrative Themes for Research on Innovation Management. *International Journal of Innovation Management* 1 (1): 1-21.

Tietz, R., J. Füller, and C. Herstatt. 2006. Signaling – An Innovative Approach to Identify Lead Users in Online Communities. In *Customer Interaction and Customer Integration*, ed. T. Blecker, G. Friedrich, L. Hvam, and K. Edwards, 453-468. Berlin: Gito.

Toffler, A. 1980. *The Third Wave*. New York: Bantam Books.

Toubia, O. and L. Florès. 2007. Adaptive Idea Screening Using Consumers. *Marketing Science* 26 (3): 342-360.

Toubia, O., J. Hauser, and R. Garcia. 2007. Probabilistic Polyhedral Methods for Adaptive Choice-Based Conjoint Analysis: Theory and Application. *Marketing Science* 26 (5): 596-610.

Tsikriktsis, N. 2005. A Review of Techniques for Treating Missing Data in OM Survey Research. *Journal of Operations Management* 24 (1): 53-62.

Tsinopoulos, C. and Z. M. F. Al-Zu'bi. 2012. Clockspeed Effectiveness of Lead Users and Product Experts. *International Journal of Operations & Production Management* 32 (9): 1097-1118.

Tuckman, B. W. and M. A. C. Jensen. 1977. Stages of Small-Group Development Revisited. *Group & Organization Management* 2 (4): 419-427.

Urban, G. L. and E. von Hippel. 1988. Lead User Analyses for the Development of New Industrial Products. *Management Science* 34 (5): 569-582.

Urban, G. L., J. R. Hauser, and J. H. Roberts. 1990. Prelaunch Forecasting of New Automobiles. *Management Science* 36 (4): 401-421.

van de Vrande, V., J. de Jong, W. Vanhaverbeke, and M. de Rochemont. 2009. Open Innovation in SMEs: Trends, Motives and Management Challenges. *Technovation* 29 (6/7): 423-437.

van Eck, P. S., W. Jager, and P. S. H. Leeflang. 2011. Opinion Leaders' Role in Innovation Diffusion: A Simulation Study. *Journal of Product Innovation Management* 28 (2): 187-203.

VDMA. 2007. *Maschinenbau in Zahl und Bild 2007*. Frankfurt am Main: VDMA Verband Deutscher Maschinen- und Anlagenbau e.V.

VDMA. 2010. *Maschinenbau in Zahl und Bild 2010*. Frankfurt am Main: VDMA Verband Deutscher Maschinen- und Anlagenbau e.V.

VDMA. 2013. *VDMA Mechanical Engineering – Figures and charts 2013*. Frankfurt am Main: VDMA Verband Deutscher Maschinen- und Anlagenbau e.V.

VDMA. 2014. *VDMA Mechanical Engineering – Figures and charts 2014*. Frankfurt am Main: VDMA Verband Deutscher Maschinen- und Anlagenbau e.V.

Verlegh, P. W. J., H. N. J. Schifferstein, and D. R. Wittink. 2002. Range and Number-of-Levels Effects in Derived and Stated Measures of Attribute Importance. *Marketing Letters* 13 (1): 41-52.

Verworn, B., C. Herstatt, and A. Nagahira. 2008. The Fuzzy Front End of Japanese New Product Development Projects: Impact on Success and Differences between Incremental and Radical Projects. *R&D Management* 38 (1): 1-19.

Voeth, M. (1999): 25 Jahre conjointanalytische Forschung in Deutschland. *Zeitschrift für Betriebswirtschaft* 2: 153-176

Volkswagen. 2009. *Driving Ideas: Annual Report 2008*. Wolfsburg: Volkswagen AG.

von Corswant, F. and P. Fredriksson. 2002. Sourcing Trends in the Car Industry: A Survey of Car Manufacturers' and Suppliers' Strategies and Relations. *International Journal of Operations & Production Management* 22 (7): 741-758.

von Hippel, E. 1976. The Dominant Role of Users in the Scientific Instrument Innovation Process. *Research Policy* 5 (3): 212-239.

von Hippel, E. 1978. Successful Industrial Products from Customer Ideas. *Journal of Marketing* 42 (1): 39-49.

von Hippel, E. 1986. Lead-Users: A Source of Novel Product Concepts. *Management Science* 32 (7): 791-805.

von Hippel, E. 1994. "Sticky Information" and the Locus of Problem Solving: Implications for Innovation. *Management Science* 40 (4): 429-439.

von Hippel, E. 2001. Perspective: User Toolkits for Innovation. *Journal of Product Innovation Management* 18 (4): 247-257.

von Hippel, E. 2005. *Democratizing Innovation*. Cambridge: The MIT Press.

von Hippel, E. and H. DeMonaco. 2013. Market Failure in the Diffusion of User Innovations: The Case of 'Off-Label' Innovations by Medical Clinicians. *Working Paper*. MIT Sloan School of Management.

von Hippel, E. and W. Riggs. 1996. A Lead User Study of Electronic Home Banking Services: Lessons from the Learning Curve. *Working Paper*. MIT Sloan School of Management.

von Hippel, E. and G. von Krogh. 2003. Open Source Software and the "Private-Collective" Innovation Model: Issues for Organization Science. *Organization Science* 14 (2): 209-223.

von Hippel, E. and G. von Krogh. 2006. Free Revealing and the Private-Collective Model for Innovation Incentives. *R&D Management* 36 (3): 295-306.

von Hippel, E. and G. von Krogh. 2013. Identifying Viable "Need-Solution Pairs": Problem Solving without Problem Formulation. *Working Paper*. MIT Sloan School of Management.

von Hippel, E., N. Franke, and R. Prügl. 2009. Pyramiding: Efficient Search for Rare Subjects. *Research Policy* 38 (9): 1397-1406.

von Hippel, E., S. Ogawa, and J. P. J. de Jong. 2011. The Age of the Consumer-Innovator. *MIT Sloan Management Review* 53 (1): 27-35.

von Hippel, E., S. Thomke, and M. Sonnack. 1999. Creating Breakthroughs at 3M. *Harvard Business Review* 77 (5): 47-57.

Wagner, P. and F. Piller. 2011. *Mit der Lead-User-Methode zum Innovationserfolg, Ein Leitfaden zur praktischen Umsetzung*. Leipzig: Handelshochschule Leipzig gGmbH.

Watts, D. J., P. S. Dodds, and M. E. Newman. 2002. Identity and Search in Social Networks. *Science* 296 (5571): 1302-1305.

Webster, F. E. and Y. J. Wind. 1972. A General Model for Understanding Organizational Buying Behaviour. *Journal of Marketing* 36 (2): 12-19.

Weiber, R. and D. Mühlhaus. 2009. Auswahl von Eigenschaften und Ausprägungen bei der Conjointanalyse. In *Conjointanalyse*, ed. D. Baier, and M. Brusch, 43-58. Berlin: Springer-Verlag.

Wendelken, A., F. Danzinger, C. Rau, and K. M. Moeslein. 2014. Innovation Without Me: Why Employees Do (Not) Participate in Organizational Innovation Communities. *R&D Management* 44 (2): 217-236.

Wessling, C. 2011. Anfällige Anlagen. *Technology Review – MIT's Magazine of Innovation* 9 (11): 72-74.

Wilkie, W. L. and E. A. Pessemier. 1973. Issues in Marketing's Use of Multi-Attribute Attitude Models. *Journal of Marketing Research* 10 (4): 428-441.

Wind, Y. J. and V. Mahajan. 1997. Issues and Opportunities in New Product Development: An Introduction to the Special Issue. *Journal of Marketing Research* 34 (1): 1-12.

Wind, Y. J. and T. L. Saaty. 1980. Marketing Applications of the Analytic Hierarchy Process. *Management Science* 26 (7): 641-658.

Wissenschaftsstatistik. 2007. *FuE-Datenreport 2007 - Tabellen und Daten*. Essen: Wissenschaftsstatistik GmbH im Stifterverband für die Deutsche Wissenschaft.

Wissenschaftsstatistik. 2009. *FuE-Datenreport 2009 - Tabellen und Daten*. Essen: Wissenschaftsstatistik GmbH im Stifterverband für die Deutsche Wissenschaft.

Wissenschaftsstatistik. 2013. *FuE-Datenreport 2013 - Analysen und Vergleiche*. Essen: Wissenschaftsstatistik GmbH im Stifterverband für die Deutsche Wissenschaft.

Witell, L., P. Kristensson, A. Gustafsson, and M. Löfgren. 2011. Idea Generation: Customer Co-Creation versus Traditional Market Research Techniques. *Working Paper*. Karlstad University.

Wittink, D. R. and P. Cattin. 1989. Commercial Use of Conjoint Analysis: An Update. *Journal of Marketing* 53 (3): 91-96.

Wittink, D. R., M. Vriens, and W. Burhenne. 1994. Commercial Use of Conjoint Analysis in Europe: Results and Critical Reflections International. *Journal of Research in Marketing* 11 (1): 41-52.

Wolter, F. and A Knie. 2011. *BeMobility - Berlin elektroMobil: Abschlussbericht*. Berlin: Deutsche Bahn AG/DB FuhrparkService GmbH.

Wright, P. and M. A. Kriewall. State-of-Mind Effects on the Accuracy with Which Utility Functions Predict Marketplace Choice. *Journal of Marketing Research* 17 (3): 277-293.

Wu, W.-Y. and B. M. Sukoco. 2010. Why Should I Share? Examining Consumers' Motives and Trust on Knowledge Sharing. *Journal of Computer Information Systems* 50 (4): 11-19.

ZEW. 2013. *ZEW Branchenreport Innovation: Fahrzeugbau*. Mannheim: Zentrum für europäische Wirtschaftsforschung GmbH.

Web-based References

Amazon. 2014b. *About Recommendations.* Available from http://www.amazon.com/gp/help/customer/display.html/ref=hp_left_v4_sib? ie=UTF8&nodeId=16465251 [accessed 15 September 2014].

Blanchet, M., T. Rinn, G. Von Thaden, and G. de Thieulloy. 2014. *INDUSTRY 4.0: The new industrial revolution: How Europe will succeed.* Available from http://www.rolandberger.com/media/pdf/Roland_Berger_TAB_Industry_4_0_ 20140403.pdf [accessed 15 June 2014].

Blankenhorn, D. 2015. *Why The Internet Of Things Matters So Much To ARM Holdings.* Available from http://seekingalpha.com/article/2898186-why-the-internet-of-things-matters-so-much-to-arm-holdings [accessed 1 March 2015].

Borchers, D. and A. Wilkens. 2012. *CAST diskutiert strukturelle Defizite kritischer Infrastrukturen.* Available from http://heise.de/-1728627 [accessed 29 April 2015].

Bosnjak, M. 2013. *Gehört Mobile Research die Zukunft? Invited talk given for marktforschung.de.* Available from http://www.prof-bosnjak.de/bosnjak/media/2013_02_18_MobileBefragungen_Bosnjak.pdf [accessed 2 November 2014].

Churchill, J., E. von Hippel, and M. Sonnack. 2009. *Lead-User Project Handbook: A Practical Guide for Lead User Project Teams.* Available from http://web.mit.edu/evhippel/www/teaching.htm [accessed 4 May 2014].

Destatis. 2014. *Kennzahlen der Unternehmen des Verarbeitenden Gewerbes 2012.* Available from https://www.destatis.de/DE/ZahlenFakten/Wirtschaftsbereiche/IndustrieVerar beitendesGewerbe/Tabellen/KennzahlenVerarbeitendesGewerbe.html [accessed 15 June 2014].

Eikenberg, R. 2015. *Vorsicht! Microsoft-Patch legt Rechner lahm.* Available from http://heise.de/-2545913 [accessed 1 March 2015].

ESCI. 2011. *ESCI – Enhanced Security for Critical Infrastructures.* Frankfurt (Oder): IHP - Innovations for High Performance Microelectronics. Available from http://www.esci-vrs.de/wp-content/uploads/2015/04/ESCI-2011-Whitepaper.pdf [accessed 30 April 2015]

ESOMAR. 2014. *Broader 'business intelligence' market surges 50% to US$ 60 billion*. Available from http://www.esomar.org/news-and-multimedia/news.php?pages=1&idnews=150 [accessed 27 October 2014].

Genesbmx. 2014. *Gene`s BMX Reviews*. Available from http://www.genesbmx.com/reviews.html [accessed 21 September 2014].

Ghosn, C. 2011. *Nissan Power 88: Delivering the Full Potential of the Company*. Available from http://www.nissan-global.com/EN/IR/MESSAGE/ [accessed 15 June 2014].

Hahsler, M. 2011. *recommenderlab: A Framework for Developing and Testing Recommendation Algorithms*. Available from http://www.icesi.edu.co/CRAN/web/packages/recommenderlab/vignettes/recommenderlab.pdf [accessed 27 October 2014].

Hahsler, M. 2014. *Package 'recommenderlab'*. Available from http://cran.r-project.org/web/packages/recommenderlab/recommenderlab.pdf [accessed 27 October 2014].

Henne, S. 2010. *Lead User Methode bei DATEV*. Available from http://wi1.uni-erlangen.de/sites/wi1.uni-erlangen.de/files/2010-11-22_gastvortrag_steffen_henne_datev.pdf [accessed 2 November 2014].

Herbst, K. 2012. *Hacks mit schweren Folgen*. Available from http://www.deutschlandfunk.de/hacks-mit-schweren-folgen.684.de.html?dram:article_id=224227 [accessed 29 April 2015].

Herkommer, G. 2013. *VDMA - Weltweiter Maschinenbau - die Prognose für 2014*. Available from http://www.computer-automation.de/steuerungsebene/steuern-regeln/artikel/102396/0/ [accessed 1 March 2015].

Hillenbrand, T. 2007. *Ford-Flop Edsel: Titanic auf Rädern*. Available from http://www.spiegel.de/einestages/ford-flop-edsel-a-948600.html [accessed 27 October 2014].

Jester. 2014. *Jokes for Your Sense of Humor.* Availabe from
  http://eigentaste.berkeley.edu/user/ [accessed 15 September 2014].

Maurice, E. P. 2013. *Maintaining the security-worthiness of Java is Oracle's
  priority.* Available from
  https://blogs.oracle.com/security/entry/maintaining_the_security_worthinew
  o_of [accessed 3 March 2015].

Meinberg, U. 2006. *Sensorsysteme in Wertschöpfungsketten.* Available from
  http://www.leibniz-institut.de/cms/pdf/Meinberg-
  Sensorsysteme_in_Wertschoepfungsnetzen.pdf [accessed 18 June 2014].

Sänn, A. and O. Stecklina. 2013. *Impuls-Vortrag zur Thematik: Sicherheitsvorfälle
  in KRITIS seit 02.09.2011.* Frankfurt am Main: AK KRITIS Workshop.
  Available from http://www.esci-vrs.de/wp-content/uploads/2013/01/AK-
  KRITIS-25.01.2013.pdf [accessed 30 April 2015].

Schmitz, C. 2012. *LimeSurvey: An Open Source Survey Tool.* Available from
  http://www.limesurvey.org/en/about-limesurvey/license [accessed 1
  November 2014].

Spaar, D. 2015. *Auto, öffne dich! Sicherheitslücken bei BMWs ConnectedDrive.*
  Available from http://heise.de/-2536384 [accessed 4 March 2015].

Volkswagen. 2010. *Launch of the Home Power Plant.* Available from
  http://www.volkswagenag.com/content/vwcorp/info_center/en/news/2010/1
  1/Launch_of_the_home_power_plant.html [accessed 15 June 2014].

Vollrath, C. 2002. *Optimierung der Hersteller-Zulieferer-Beziehung durch
  „Networked" Supply Chain Management.* Available from
  http://www.competence-site.de/pps-systeme/Optimierung-Hersteller-
  Zulieferer-Beziehung-durch-Networked-Supply-Chain-Management
  [accessed 15 June 2014].

Printed in the United States
By Bookmasters